THE NEXT SEVENTY YEARS

POPULATION, FOOD AND RESOURCES

THE NEXT SEVENTY YEARS

POPULATION, FOOD AND RESOURCES

BERNARD GILLAND, B.Sc.

ABACUS PRESS

First Published in 1979 by Abacus Press,
Abacus House, Speldhurst Road, Tunbridge Wells, Kent, TN4 OHU

British Library Cataloguing in Publication Data

Gilland, B.
The next seventy years
1. Population 2. Natural resources
1. Title
301.32 HB8 71
ISBN 0-85626-176-9

Printed in England
by
The Garden City Press Limited

CONTENTS

FOREWORD

I thank all who responded to my enquiries in connection with the present work. I am especially grateful to Prof. Sigurd Andersen, of the Agricultural University of Denmark who read an earlier draft of the manuscript and made many helpful suggestions; to Prof. M. S. Losowsky, of the University of Leeds, for discussions on a number of problems; to Prof. R. S. Loomis, of the University of California, for his suggestions concerning photosynthetic potential; and to Prof. D. Pimentel, of Cornell University, who read and commented on the penultimate draft. They are in no way responsible for the views expressed in the book or for any errors it may contain.

I thank Mr. Erik Andersen for the fine job he did in preparing the diagrams for the press.

Espergaerde, Denmark
March 1979

Bernard Gilland

(vii)

The so-frequently expressed blind confidence in the forthcoming of adequate sustenance for human populations, no matter how fast they multiply, represents a regression to an infantile level of thought and feeling .... Any attempt to overcome this confidence and to make an unbiased scrutiny of the actual relationship between mouths and food ... [is] being opposed with an energy that derives its force from a dim realization of the fact that such a scrutiny would threaten with destruction the pious and pleasing illusions which have been built up around this subject.

J. C. Flugel, "Men and Their Motives"
New York, 1947

I
INTRODUCTION

THE PROBLEM

The world food-population problem has two aspects. The first is that a quarter of the world's population (the inhabitants of the developed countries) enjoy diets more than adequate in energy and protein, and in which over half the protein is derived from meat, milk, eggs and fish, while the remaining three quarters (the inhabitants of the less-developed countries) subsist on diets which, on average, are inadequate in energy and protein, and in which by far the greater part of the protein is derived from cereals, pulses and starchy roots. The question is whether these disparities can be eliminated.

The second is that whereas the population of the developed countries is likely to stabilize at a level approximately 50% above the present figure, that of the less-developed countries is certain to increase at least two and a half times, and may increase fourfold, before stabilization is achieved. The question is whether per capita food supply in these countries can be maintained at least at present levels until the population stabilizes.

DEVELOPED AND LESS DEVELOPED COUNTRIES

The developed countries (hereafter referred to as DCs) will be taken to comprise USA, Canada, all European countries, USSR, Japan, Australia and New Zealand, the rest of the world constituting the less-developed countries (LDCs). Some of the data presented in this book are based on the FAO classification, in which South Africa and Israel are also included in the category of developed countries.

The DC category has not received any new members during the past two decades. There are several countries in a transition phase, notably Taiwan, South Korea and Singapore, and which may be expected to attain DC status before long. The countries of temperate South America, particularly Argentina and Uruguay, are borderline cases, which have been hovering on the brink of DC status for many years; whether and when they will join the ranks of the DCs is problematic.

THE ORIGIN OF THE DC–LDC DICHOTOMY

It has been alleged that the DCs, or at least the West European DCs, are in large measure responsible for the existence of the global development gap. Dr. K. Singh expressed this view at the World Population Conference at Bucharest in August 1974:

> The dichotomy between the rich and the poor countries must be viewed in the broad historical perspective of the last several centuries. The departure from a certain uniform level came with the industrial revolution, a process in which both the presently developed and the developing countries have participated.
>
> The developed countries have been the beneficiaries of industrial progress, to which the developing countries were compelled to contribute substantially through a process of colonial denudation over several centuries. This process of exploitation consolidated certain untoward relationships, both political and economic. As a result of this the positive development processes which the underdeveloped world has been trying to work out over the last quarter of a century are being pursued under severe constraints. The developed countries are pushing ahead rapidly with the lead they have gained in the fields of technology and economic growth, and the gap between the two worlds is widening. This situation has to change.

Dr. Singh's first postulate, that there was no development gap on the eve of the industrial revolution, is simply untrue – the per capita product of N.W. Europe about the year 1800 was substantially higher than those of Asia and Africa. Nor is it true that the per capita product of the less-developed regions declined during the period 1800-1950 because of "colonial denudation" or from any other cause; on the contrary, it increased in some of the countries concerned, and remained roughly constant in the others.* It is true that the developed countries have been the principal beneficiaries of industrial progress, but this is hardly surprising, seeing that they were its originators and the driving force behind it.

The fallacies propounded by Dr. Singh are moderate, however, when compared with the statements made by others who have allowed themselves to be completely carried away by their enthusiasm for knocking the West. For example, N. Keyfitz has claimed that the development of industrial capitalism in Western nations in the period 1800-1950 resulted in the development of a one billion excess world population, largely in the tropics, as a result of exploitation of these areas for raw materials (with the resultant need for labor) during the period of colonialism. (From an article in "Journal of Social Issues" Vol. 23, January 1967.) The fantastic absurdity of Keyfitz's statement becomes clear as soon as one looks at the facts. The rate of increase of population in Europe and the areas of European settlement exceeded the rate of increase of population in the tropics throughout the period 1800-1950. Population growth in Africa and Asia during this period averaged 0.5% per year, a modest rate by any standards, and lower than the present rate in the USA, USSR and Japan. Since 1950 it has increased very rapidly, the present rate of about 2.4% per year exceeding anything ever experienced in the DCs, with the exception of the period of mass immigration in N. America and Australia.

Professor B. Commoner has alleged that "the industrial nations were able to achieve their demographic transition as a result of wealth obtained in no small measure through colonial exploitation of the very nations which are now experiencing intense population pressures. If these nations had been able to retain their resources, it is likely that they would be significantly closer to the economic conditions necessary for falling birthrates." (From a contribution to the symposium "Is there an Optimum Level of Population?" McGraw-Hill, 1971). Neither of Commoner's propositions will bear examination. It is true that Western Europe imported agricultural products such as tea, coffee, cocoa, fruit, sugar, spices, cotton and jute from the colonies in exchange for manufactured goods and bullion, but it is difficult to see how this

*Cf. S. Kuznets, "Population, Capital and Growth" New York, 1973, pp. 31 and 293.

can be interpreted as "exploitation." It is frequently alleged in this connection that the terms of trade were "unduly" favourable to the industrial countries. Leaving aside the question of the meaning of "unduly" in this context, the fact is that even if the terms of trade had been more favourable to the colonies they could not have sufficed to start the industrial revolution in those countries. The primary characteristic of industrial development throughout the nineteenth century was rapidly increasing production and utilization of coal and iron, a process which required accessible resources, technological inventiveness, and a class of administrators, all of which were lacking in Africa and South Asia. If it be suggested that the Western European countries themselves should have started the colonies on the road to industrialization, one can only reply that such a policy was impossible with the resources then available. The European countries could not have acted otherwise than they did, that is, by developing their own resources first. As soon as the phase of massive capital accumulation was completed, demands for higher living standards began to make themselves felt. By the time large-scale capital investment in Africa and Asia became economically possible (with the advent of the "affluent society") almost all the colonies had achieved independence, and the ensuing political conditions were not such as to encourage investment by the developed countries.

Even if Commoner's argument had any validity, it would apply only to the former colonial powers (Britain, France, Netherlands, Belgium, Spain, Portugal) which together have less than one sixth of the population of the developed countries.

Commoner's second proposition implies that if the former colonies had retained their tea, cotton etc. instead of exporting these products to Europe, they would now be much more advanced economically. Recent experience shows that if anything the reverse is true; the higher the volume of foreign trade relative to national product, the more rapid is the growth of national product. Commoner is apparently also unaware that those LDCs which have never been colonies of European powers – China, Thailand, Iran and Turkey among others – were until recently no more advanced industrially than those which have. Finally, it is not at all clear that any specific set of economic conditions are necessary for falling birth rates. Though there is a rough inverse correlation between per capita product and birth rate, it is nowhere near high enough to support Commoner's assumption. This problem will be discussed in Chapter II.

In view of the current tendency to over-emphasize political and economic factors as causes of the development dichotomy, it will be well to note a theory which goes to the opposite extreme.

The American geographer Ellsworth Huntington (1876-1947) in his book "Civilization and Climate" (Yale University Press, 1915; 3rd ed. 1925, reprinted in 1971 by Shoe String Press, Hamden, Connecticut) developed a classification of regions based on what he called climatic energy. Huntington held that a mean temperature of 10°C, a wide seasonal temperature variation and a high frequency of atmospheric pressure minima or humidity reversals were most conducive to mental and physical efficiency. He computed a climatic energy index for the various regions, the index being a function of temperature and pressure variations. The result of his calculations can be summarized as follows: Regions of high climatic energy are USA (excepting the South and the South-Western desert area), southern Canada, Europe (excepting southern Iberia, Sicily, Greece and Caucasia), western Siberia south of latitude 60°, Japan, Korea, south-eastern Australia, Tasmania and New Zealand. Regions of very low climatic energy are the Amazon and Orinoco basins, the Caribbean, north-east Brazil, the Sahara and Sahel, West Africa, the Congo basin, Arabia, southern Iran, the Indian sub-continent and South-East Asia. The remaining regions of the world have low to medium climatic energy. The correlation between degree of economic development and climatic energy index is even more striking today than when Huntington's book appeared.

Though Huntington's climatic energy theory is unproven, Dr. A. M. Kamarck (in "The Tropics and Economic Development", Johns Hopkins University Press, Baltimore, 1976) has suggested that climate is an indirect cause of the backwardness of the tropical regions, insofar as the absence of an annual cold period favours the persistence of a number of endemic diseases and pests which attack plants, animals, and humans.

It is unnecessary for the present purpose to delve more deeply into the causes of the global development dichotomy. It is clear, however, that there is no justification for blaming any specific group of countries for its existence.

THE FUTURE OF THE DICHOTOMY

There has been much speculation on the long-term prospects for global economic development. There is general agreement that the dichotomy will sooner or later disappear, but vast differences of opinion regarding when and how this will take place. There are the technological optimists, such as Herman Kahn, who believe that the LDCs will catch up with the DCs within a couple of centuries or so even if exponential growth of per capita product in the latter continues indefinitely, and that the inhabitants of all regions will thereafter enjoy high and steadily rising material standards. Some go even further, and hold that continued economic growth in the DCs is a necessary condition for continued growth in the LDCs. On the other hand, there are many who believe that if an approximate uniformity in economic development among the world's regions is ever to be achieved, it will have to be at a level not exceeding, and possibly well below, the current level in the most advanced industrial countries. A third view, held by Charles Galton Darwin, is that technological civilization is destined to disappear, and that all regions will ultimately return to the pre-industrial state, with economies based on subsistence agriculture. Speculations of this nature lie outside the scope of the present study, which is chiefly concerned with the global demographic, agricultural and energy situation at the present time and the outlook for the period up to the year 2050, when world population is expected to be close to stabilization. Certain ecological problems arising from increasing food and energy production will be touched on, but no attempt will be made to deal with other problems (urbanization, education, employment, depletion of resources other than fossil fuels, etc) created by continuing growth of population and production.

II
POPULATION

Since the end of the Second World War the annual rate of population increase in the developed countries has been slowly declining, while that in the less-developed countries has been steadily rising. It is only since the late nineteen sixties, however, that the long-range implications of the "population explosion" in the LDCs have become apparent. In the past few years several global demographic projections have been computed. Before reviewing two of these projections, it is necessary to briefly discuss the cause of current population growth and describe certain basic demographic concepts.

The primary cause of the rapid population increase in the LDCs is declining mortality, which in turn has been caused by the achievements of modern medicine, particularly antibiotics and serums. Birth-rates have either remained constant or declined very slowly in most LDCs over the past thirty years, while death-rates have plummeted. The result is an extremely rapid annual increase of population, an increase which in the limit can exceed 4%. The current average rate of population growth in the LDCs is 2.4% per year, in the DCs 0.7%.

If birth- and death-rates in the LDCs were to remain at present levels, the population of these regions would double every thirty years, reaching almost 30 billion by 2075. There is general agreement that such a development is both highly undesirable and extremely unlikely to happen. There are, however, only two ways in which it can be prevented: declining birth-rates or rising death-rates. A combination of both is theoretically possible, but very improbable, as rising mortality would almost certainly result in rising fertility.

Long-range demographic projections have so far been based on the assumption of declining fertility, the main difference between the various projections being the rate at which fertility is assumed to decline. This is not because rising mortality is unlikely, but because the recent mortality decline in the LDCs is continuing, and any assumption regarding the time of onset of a turning point and the subsequent rate of mortality increase would be purely arbitrary.

The measures of fertility used in demographic projections are the gross reproduction rate (GRR) and the net reproduction rate (NRR). The GRR is the number of live-born daughters who would be born to 1000 females as they pass through the childbearing period (from the age of 15 to 49) according to the age-specific maternity rates existing in a given year. The NRR is the number of live-born daughters who would be born to 1000 females during the period from birth to the end of the childbearing period (from the age of 0 to 49) according to the age-specific maternity rates and mortality rates in a given year. The GRR is purely a measure of fertility, and the NRR a measure of fertility corrected for mortality. The ratio of NRR to GRR is the proportion of live-born females who survive to the mean age of childbearing. The method by which the two reproduction rates are calculated is shown in Table I. The data given in the Table show that a cohort of 1000 women in a typical DC in 1965-70 gave birth to an average of 37.7 daughters each year during the first five years of the childbearing period, 71.5 daughters per year during the second five years, and so on. The GRR is therefore five times the sum of the age-specific maternity rates, or 1300. As 96% of females survived to the mean age of childbearing, it follows that the NRR is 0.96 × 1300 = 1250.

Figure I shows the UN estimates for GRR and NRR in 22 regions in the period 1970-5. Figure 2 shows the mean expectation of life at birth in the same regions. Figure 3 shows the age distribution of the population in each region. If the NRR of a population remained constant at 1000, and the expectation of life constant at 74.8 years, the age distribution of the population would ultimately stabilize with the 0-14

Table 1.

		DC	LDC
Age specific maternity rates (1965–70), per thousand			
Age group (years):	15–19	37.7	50.4
	20–24	71.5	117.6
	25–29	70.2	156.8
	30–34	48.1	123.2
	35–39	26.0	78.4
	40–44	5.2	28.0
	45–49	1.3	5.6
	Total	260.0	560.0
Gross reproduction rate		1300	2800
Mean age of child bearing		26.5 y	28.5 y
Proportion of females surviving to mean age of childbearing		0.96	0.74
Net reproduction rate		1250	2070

Source: T. Frejka, "The Future of Population Growth", 1973.

years group constituting 19.7% of the population, and those aged 65 years and over constituting 16.9%.

THE UN DEMOGRAPHIC PROJECTIONS

The UN population projections for the period 1975-2075, and the assumptions on which they are based, are given in "Concise Report on the World Population Situation in 1970-1975 and its Long-Range Implications," published by the UN in 1974 as Population Study No. 56.

For the purpose of the projection the world was divided into eight "major areas": North America (USA and Canada), Europe, USSR, East Asia, Latin America, Africa, South Asia and Oceania. The projection was computed on the assumption that the net reproduction rate in each of the eight areas will decline linearly to 1000 and thereafter remain constant. It was also assumed that when the net reproduction rate reaches 1000, the gross reproduction rate will be 1015, i.e. that 98.5%

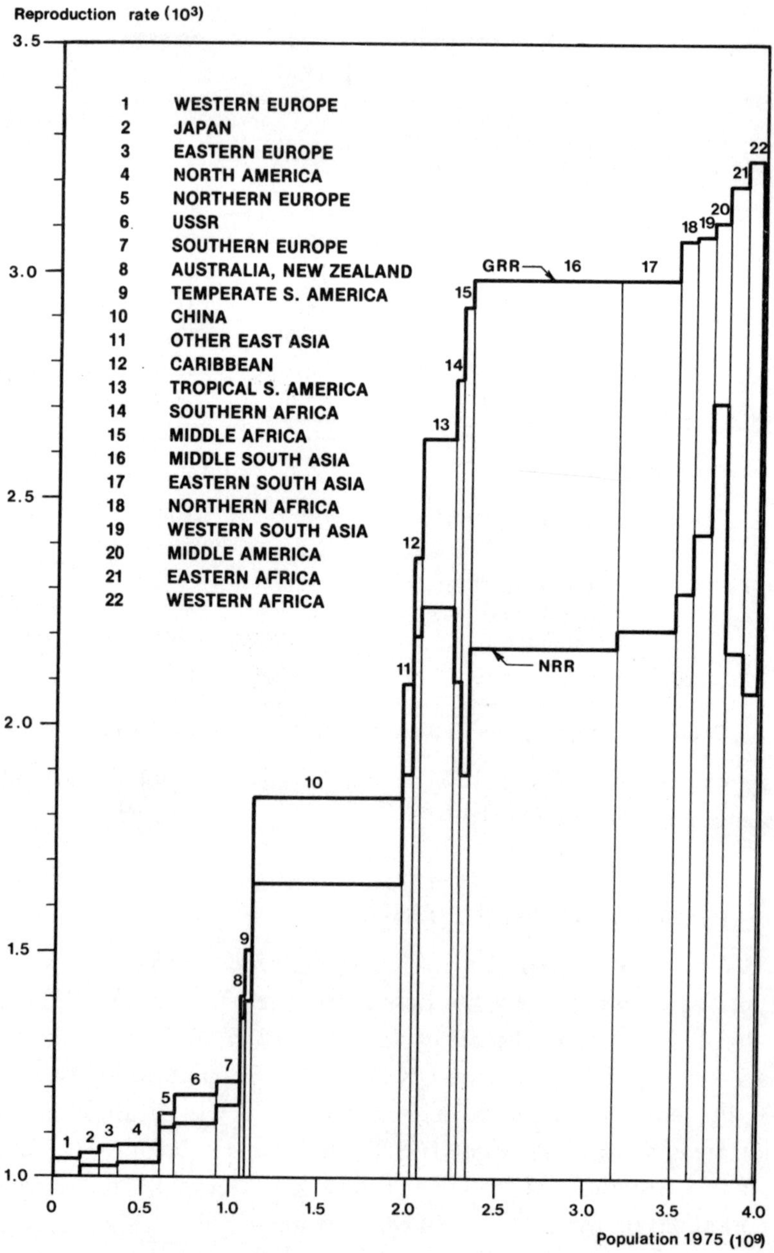

Fig. 1. Gross and net reproduction rates 1970–75. (Source: UN Population Study No. 56).

of liveborn females will survive to the mean age of childbearing. The net reproduction rate in each area is assumed to reach 1000 in the year shown in Table 2. The projection computed on these assumptions is known as the "medium" projection. To take account of the uncertainty of the date at which the net reproduction rate reaches 1000, a "high" and a "low" projection were also computed. For the "high" variant the date is assumed to be ten years later, and for the "low" variant ten years earlier, than the dates given in Table 2.

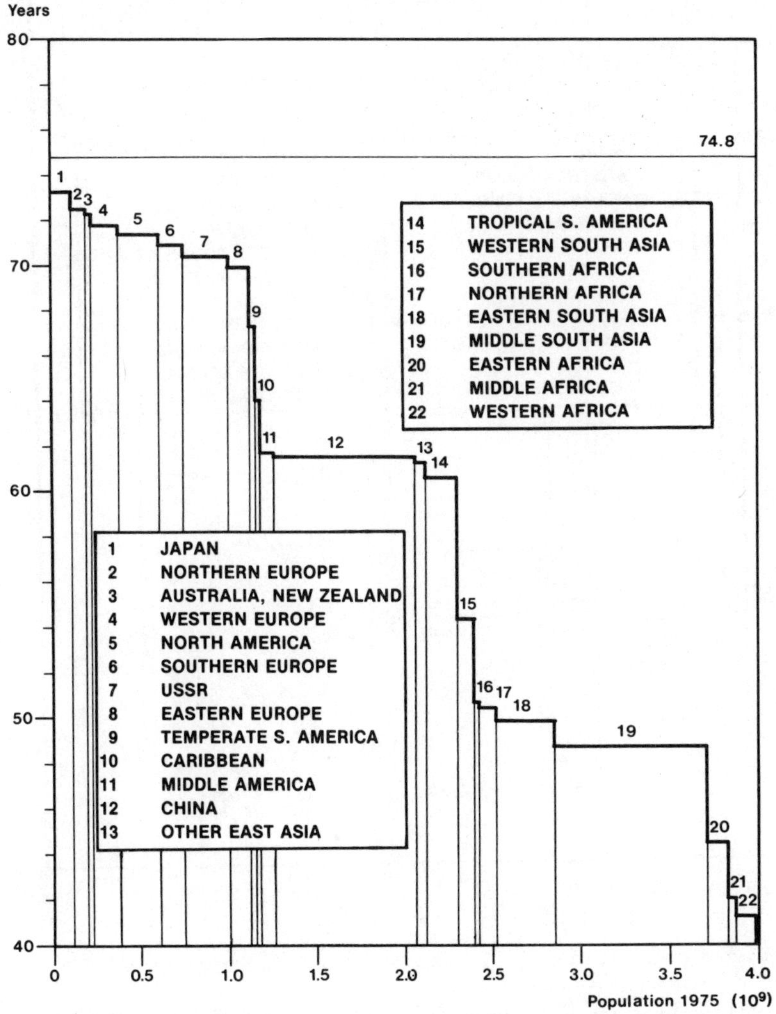

Fig. 2. Expectation of life at birth, 1970-75. (Source: UN Population Study No. 56).

The projections are also based on the assumption that the expectation of life at birth in each major area will rise linearly to 74.8 years,* and thereafter remain constant. For all projections it is assumed that this expectation of life will be attained in each area in the year indicated in Table 2.

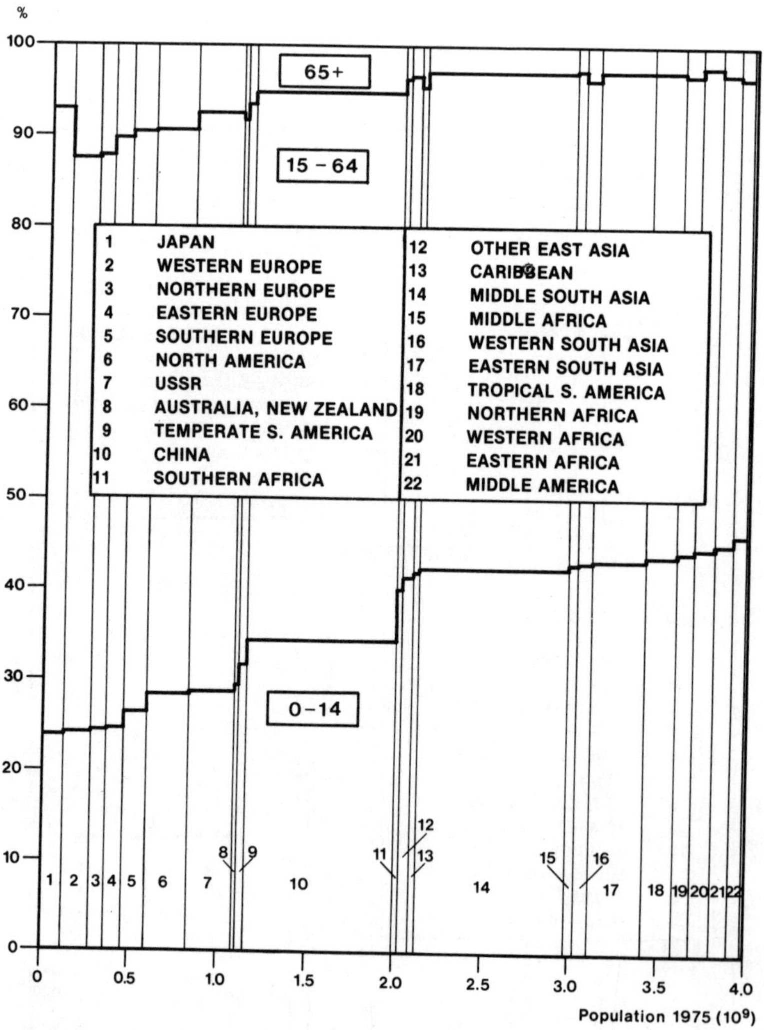

Fig. 3. Age distribution, 1970. (Source: UN Population Study No. 56).

*The mean expectation. The projected expectation for females is 77.5 years, for males 72.5 years.

In addition to the "high", "medium" and "low" projections, an "instant fertility reduction" variant was computed, for which it is assumed that the net reproduction rate fell to 1000 in 1970. This projection is of interest as it shows the momentum of population growth in each major area, i.e. the future population increase due solely to the present age distribution and the assumed future increase in expectation of life.

The population in 1975 and the projected population in 2075 of each of the eight major areas on each of the four projections is given in Table 3.

Table 2.

Major area	Year in which NRR = 1000	Year in which expectation of life at birth = 74.8 yrs
Europe	2005	2015
North America	2010	2030
USSR	2015	2025
Oceania	2020	2035
East Asia	2020	2040
Latin America	2035	2040
South Asia	2060	2070
Africa	2070	2075

An examination of Table 3 shows inconsistencies and discrepancies which do not inspire confidence in the competence of the UN demographers. These weaknesses will become clearer when the UN projections are compared with those computed by T. Frejka. Before doing so, the UN figures can be criticized on the following counts:

1. The projected populations of North America and Europe on the "low" variant are lower than those on the "instant fertility reduction" variant. Now in the "low" variant it is assumed that the NRR falls to 1000 (i.e. replacement level) ten years earlier than is assumed in the "medium" projection. Reference to Table 2 shows that the NRRs in North America and Europe on the "low" projection reach 1000 in

Table 3.

Area	1975	2075 (Instant fertility reduction)	2075 (Low)	2075 (Medium)	2075 (High)
	Population (millions)				
North America	243	311	295	340	488
Europe	479	552	533	592	669
USSR	256	326	359	400	435
Oceania	22	28	40	52	66
Latin America	327	495	1003	1297	1796
Africa	395	617	1599	2522	3465
South Asia	1296	1881	4102	5232	6898
East Asia	1011	1405	1531	1775	2014
World	4029	5615	9462	12210	15831

the years 2000 and 1995 respectively. The "instant fertility reduction" variant, on the other hand, is based on the assumption that NRR fell from its actual level to 1000 in 1970. If all other assumptions are the same, the "instant fertility reduction" assumption *must* result in a lower population at any time after 1970 than the population which would result from a slower fertility decline. It is certain, then, either that the figures for the "low" variant for the two areas in question are incorrect, or that the figures for the "instant fertility reduction" variant are incorrect.

2. The projected population of East Asia on the "low" variant is only 9% greater than on the "instant fertility reduction" variant. As the NRR in East Asia on the "low" variant is supposed to reach 1000 in the year 2010, it is incredible that a difference of 40 years in the date at which NRR reaches 1000 should give rise to so small a difference, bearing in mind that the NRR in East Asia is estimated (in the UN study) at approximately 1600 in 1970.
3. The "medium" and "high" variants for East Asia are also suspect. On the "medium" projection a fall in the NRR from 1600 to 1000 by 2020 results in a 76% increase of population in 2075 relative to that in 1975. Yet on the "medium" projection for North America, for example, a fall in the NRR from 1030 to 1000 by the year 2010 in a

much "older" population results in an increase of 40% relative to the 1975 population. Either the projected population of East Asia is too low or that of North America is too high. This criticism applies with even greater force to the "high" projection, in which the relative increase in North America is slightly greater than that in East Asia!

T. FREJKA'S PROJECTIONS

Dr. T. Frejka computed five demographic projections for each of the eight major areas. The projections are given in his book, "The Future of Population Growth," (New York, 1973). The projections are based on mortality assumptions similar to those of the UN projections – female expectation of life at birth rising to 77.5 years in the DCs and 74 years in the LDCs by 2045-50. The fertility assumptions are as follows:

Projection	NRR declines to 1000 in the period
1	1970–75
2	1980–85
3	2000–05
4	2020–25
5	2040–45

Table 4.

Area	Population in 2050 (millions) Proj. 1	Proj. 2	Proj. 3	Proj. 4	Proj. 5	UN 'medium' proj.
North America	312	329	368	406	444	339
Europe	565	589	641	692	739	592
USSR	327	335	353	371	387	393
Oceania	27	29	35	41	47	50
Latin America	494	575	797	1077	1396	1202
Africa	565	649	899	1244	1664	2112
South Asia	1854	2163	3077	4350	5916	4715
East Asia	1448	1605	2002	2487	3011	1760
World	5592	6274	8172	10668	13604	11163

Frejka s projections for the year 2050 are given in Table 4, which also shows the UN "medium" projection for that year.

A comparison of the UN projection with those of Frejka reveals a number of discrepancies. Bearing in mind the years in which NRR is assumed to reach 1000 in Frejka's five projections and in the several areas in the UN "medium" projection, it can be seen that in all areas except Latin America and Africa there is a long interval between the year in which NRR falls to 1000 on the UN projection and the year in which it would have to reach 1000 on Frejka's projection in order for the latter to give the same projected population as the UN projection. The UN projection for North America, for example, is very close to Frejka's Projection 2, though in the UN projection it is assumed that NRR reaches 1000 in the year 2010, while in Projection 2 it reaches 1000 in the period 1980-85. For the USSR, on the other hand, the UN projection is very close to Projection 5, though the former is based on the assumption that NRR falls to 1000 in the year 2015, the latter in the period 2040-45.

As the discrepancies between the UN projection and those of Frejka are not systematic, it is very likely that one or the other is computed incorrectly. We have seen, however, that the UN projections are mutually inconsistent. It is therefore reasonable to place more confidence in Frejka's projections.*

For the purposes of the present study the eight major areas are converted to five regions. North America, Europe, USSR, Oceania and Japan are considered as forming one region – the Developed Countries (DCs). China (including Taiwan), North Korea, South Korea and Mongolia form the East Asia region. The regions Latin America, Africa and South Asia are identical with the major areas of the same name.

Assuming that Frejka's projections are correctly computed on the given assumptions, the problem then arises as to which, if any, of the five assumed rates of fertility decline is most probable in each region. The fertility situation and outlook in the regions will now be considered.

*Since the publication of UN Population Study No. 56, the UN demographers have revised their projections. The revised projections, and the assumptions on which they are based, are given in UN Population Study No. 60 (1977); the new projections do not, however, go beyond the year 2000. (See Appendix H.)

DC: The NRR in North America has been declining for some years, and dropped below 1000 in the early seventies. The NRR in USA in 1974 was 879; in Canada in 1973, 911. In the USSR NRR has been almost constant in recent years, the latest available figure being 1128 in 1973.

In Japan the NRR has been fluctuating for over a decade, reaching a low of 920 in 1960, rising to 1020 in 1971 and thereafter declining to 910 in 1975.

All the countries of Northern and North-Central Europe now have NRRs under 1000, the lowest figure being that in the German Federal Republic – 710 in 1974. The German Democratic Republic is not far behind with an NRR of 732 in the same year. (If the NRRs in these two countries remain at present levels, their populations will decline by over 50% within a century).

In the remaining DCs NRR is in the range 1000-1300, and either stable or slowly declining.

What conclusion can be drawn regarding the long-term tendency of NRR? Only that, like the stock market, it will fluctuate. Future population growth could lie anywhere between Projection 1 and 5. All things considered, Projection 3 is perhaps slightly more probable than any other. It will be assumed that Projection 2 represents "minimum" and Projection 4 "maximum" growth in the period up to the year 2050.

Latin America, Africa, South Asia: As the NRR in each of these regions is at present in the range 2100-2300, and there is no reliable indication of a fertility decline in any of the most populous countries, Projections 1, 2 and 3 are out of the question as possible lines of demographic development. *Any* assumed rate of fertility decline is little more than an act of faith, but if equilibrium is to be achieved through fertility decline at all, the rate of decline cannot be expected to be more rapid than that assumed in Projection 4, while a rate of decline slower than that assumed in Projection 5 will almost certainly result in population stabilizing through rising mortality. These two projections will therefore be taken to be the "minimum" and "maximum" projections respectively.

East Asia: The demographic outlook for this region differs from that for the other LDC regions in two ways. Firstly, the present NRR is significantly lower – about 1700. Secondly, by far the greater part of the population of the region lives under Communist governments, which may be able to carry out a more effective demographic policy. In addition, the two non-Communist countries of the region (South Korea and Taiwan) are already in the transition to DC status, a transition accompanied by appreciable fertility decline. It would seem reasonable to adopt Projections 3 and 4 as the "minimum" and "maximum projections.

The "minimum" and "maximum" population projections for 2050 are given in Table 5. Japan has been included in the DC region on the assumption that the population in 2050 will be 138m on Projections 1 and 2, 139m on Projection 3, 140m on Projection 4 and 141m on Projection 5, as estimated by Frejka.

Table 5.

	Population in 2050 (millions)	
	Minimum	Maximum
DC	1420	1650
Latin America	1077	1396
Africa	1244	1664
South Asia	4350	5916
East Asia	1863	2347
World	9954	12973

THE DEMOGRAPHIC TRANSITION

Both the UN and Frejka's projections are based on the theory of the demographic transition, in which a causal connection is postulated between fertility decline and a cluster of socio-economic variables – mortality decline, education, urbanization, industrialization and enhancement of the status of women. Though there is a fair degree of correlation between fertility and such development indicators as infantile mortality, life expectancy, literacy and proportion of the labour force engaged in agriculture, little success has been achieved in correlating the timing of onset and the rate of progress of fertility decline with these variables, as M.S. Teitelbaum pointed out in "Relevance of Demographic Transition Theory for Developing Countries," Science 188, 1975. Two countries which deviate from the general trend, but in opposite directions, are Kuwait and Mauritius. The demographic transition in Kuwait has not yet begun, that country having the world's highest NRR (3260 in 1973) in spite of the per capita GDP in 1974 being 11852 dollars and the NRR/GRR ratio 0.93; in Mauritius the NRR fell from 2336 to 1525 in the period 1966-74 in spite of the per capita GDP (1974) being only 641 dollars and the NRR/GRR ratio only 0.89. Per capita GDP is admittedly only a very rough indicator of economic development, but it would be difficult to find a better health indicator than NRR/GRR. It is clear that other factors are involved in initiating and maintaining a demographic transition, factors which are not quantifiable and the effect of which cannot be accurately assessed. Yet in computing their long-range projections demographers have assumed a specific rate of fertility decline for each area, and have also assumed that the decline will continue until fertility falls to replacement level. These assumptions rest on very shaky foundations indeed. It would be a mistake, however, to dismiss long-range demographic projections as an arithmetical exercise. They should be considered as a limiting case, based on the extremely optimistic assumption that the two-child family will become the norm in all regions within the next sixty years or so. The opposite limiting case is the assumption of constant fertility in the LDCs, the death-rate ultimately equalizing with the birth-rate at 40 to 45 per thousand. This case can be called the "demographic interlude" theory, the interlude being a period in which death-rates plunge due to improved health services and subsequently rise to their original levels or higher due to mass starvation or diseases caused by malnutrition. The two cases are illustrated schematically in Figure 4.

POPULATION POLICY

There has been much talk of accelerating the demographic transition in the LDCs by creating incentives to reduce fertility. Up to the present time such measures have been largely confined to propaganda emphasizing the advantages of small families, dissemination of information on contraceptive techniques, and the sale of contraceptives at subsidized prices. Their effect on population growth has been negligible. It is highly problematic whether an effective population policy in the LDCs is at all

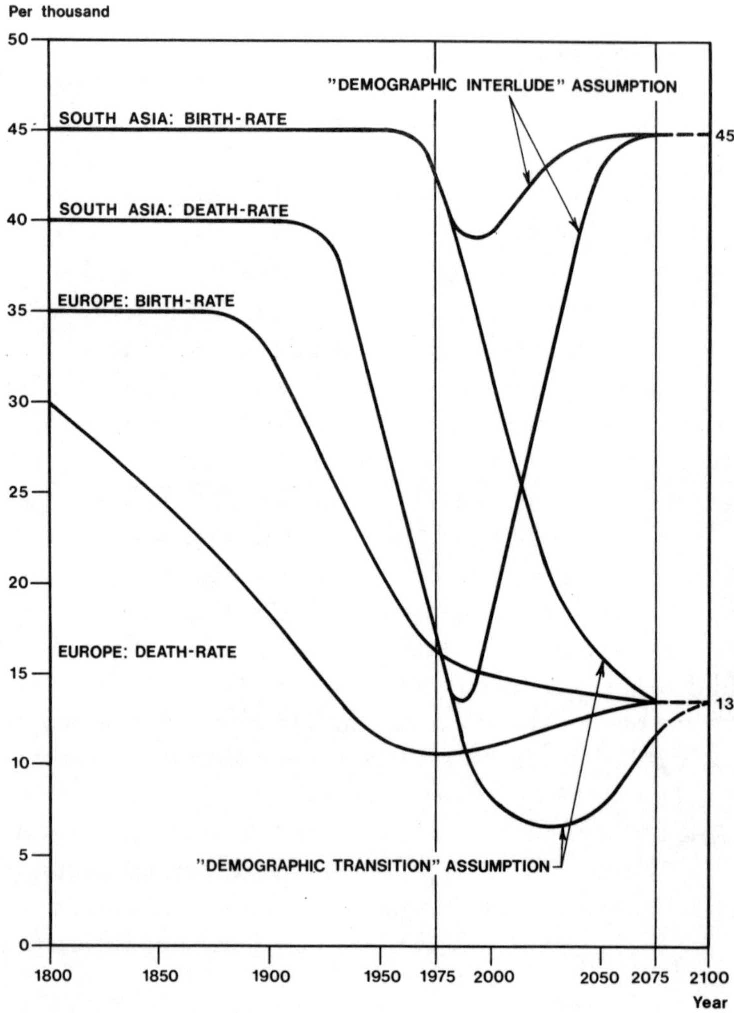

Fig. 4. Birth and death rates in Europe and South Asia, 1800–2075.

possible. The central problem is to give the great majority of parents sufficient motivation to limit their families to two children. The principal motivation operative in the DCs – a high standard of living – is impossible in the LDCs, where a living standard approaching that of the DCs is unattainable until many years (perhaps generations) after the population has stabilized. And the population of an LDC cannot stabilize by means of fertility decline until at least a half-century after the two-child family has become the norm. It is therefore necessary to create new inducements for small families if the vicious circle is to be broken. Such inducements could in theory either take the form of a comprehensive social security scheme, or of substantial economic benefits to those who limit the size of their families.

An adequate social security scheme would have to include retirement pensions, invalid pensions and unemployment assistance. Payments of one category or another would have to be disbursed to at least 10% of the population, and the cost would run to at least 5% of GNP, and very probably more. Allocations of this magnitude are out of the question for almost all the LDCs of Africa and Asia (with the exception of certain OPEC countries) and for many in Latin America. Nor can there be any question of financing social security schemes in the LDCs by aid from the DCs, as the latter cannot guarantee to disburse for an indefinite period the immense sums which would be required.

The limitation to direct incentives for small families in the LDCs is that no matter how high the scale of payments, it is very unlikely to motivate more than 25% of parents of two-child families to abstain from having a third child. Because of the patriarchal structure of society which exists in practically all LDCs, a two-child limit would only be acceptable to parents who have two *sons*; the others would wait until their second son was born before ceasing to enlarge their families, so that the average number of children per family would be not less than four. Fertility would thus remain far above the level which is necessary if population stabilization at a low level of mortality is to be achieved.

The most massive and longest sustained attempt to slow population growth by "direct" methods was the Indian sterilization programme, which started in 1956 and culminated in the carrying out of about 8 million vasectomies in 1976. The incentive was monetary compensation, the scale of payments being (from 1st May 1976) 150 rupees (approx.

18 dollars) if performed with two living children or less, 100 rupees if performed with three living children, and 70 rupees if performed with four or more children. The programme suffered from two serious defects: Firstly, the scale of payments was so low that it could act as an incentive only to those in the lowest income groups. Secondly, the medical resources, both as regards personnel and equipment, were utterly inadequate. This resulted in many cases of post-operational complications and infections. It is certain, however, that even if the incentive had been much higher and the medical resources up to European standards, the programme could not have had more than a marginal effect on India's demographic development.

DEMOGRAPHIC TRANSITION OR DEMOGRAPHIC INTERLUDE?

Whether the demographic transition or the demographic interlude will prove closest to the future course of events in the LDCs depends on whether or not per capita food production in these countries can be maintained at least at present levels until the population stabilizes due to spontaneous fertility decline – always assuming that there will be a continuous fertility decline.

In Chapter VI the maximum production of cereal grains attainable in each region by the year 2050 will be estimated. By combining these estimates with the "maximum" and "minimum" population projections, estimates for per capita grain production are made. Charles Galton Darwin, in "The Next Million Years," (London, 1952) stated that "the forecast of the future numbers of mankind is the same thing as the forecast of the future of agriculture." This is true if all or most of mankind will be living at subsistence level on a vegetarian or quasi-vegetarian diet, as Darwin believed. Darwin, incidentally, writing at a time when world population was 2500 million, estimated that "the population of the world is never likely to be more than about three to five times its present numbers." This accords fairly well with the projections of Table 5, as population growth after 2050 is relatively modest on all projections.

Darwin argued that it is impossible for world population to stabilize by means of spontaneous fertility decline, that a self-imposed national population limit would inevitably be overstepped sooner or later, and that a global demographic equilibrium established and maintained by international agreement would be inherently unstable and therefore doomed to break down. His first argument is as follows. World-wide population stabilization by spontaneous fertility decline cannot occur "because everything happens in the wrong order. What is required is the simultaneous existence of high prosperity, social conditions in which the economic disadvantages of parenthood are evident to nearly all classes, and some knowledge of the methods of birth-control." Increasing security of life results in a higher rate of population growth, which in turn keeps the standard of living down and prevents the achievement of high prosperity. As Darwin puts it, "the first condition needed for spontaneous limitation itself destroys the chances of establishing the other conditions." In reply to the objection that attainment of prosperity has been followed by fertility decline in the developed countries, Darwin points out that "the present era has been unique in that it has combined the wonders of the scientific revolution with the sudden expansion of the white races into vast almost uninhabited regions. The consequence has been that for two or three generations the Malthusian threat did really disappear. In spite of the secure conditions, man could not breed fast enough to catch up with the extending agriculture, and so the other conditions for spontaneous limitation could come into play, before the first one, the condition of security, had killed the chance of them. It would seem unlikely that similar conditions can arise again in world history, so that in estimating future possibilities, there is little prospect that the Malthusian threat will again be overcome spontaneously in this way." This argument has great cogency, and shows how weak are the assumptions made by the demographers and the demographic transition theorists.

Darwin's second argument is that a self-imposed national population limit must sooner or later be overstepped as a result of the emergence of a pronatalist political movement or religious sect, while an international agreement on population limitation (in the exceedingly unlikely event of it coming about) would break down because in the last resort the only way to enforce it would be war, and it is certain that most countries would be very reluctant to resort to war solely in order to compel a country which had violated the agreement to reduced its population. The only answer which can be made to this argument is that considerations of what might happen, or even of what must happen, in the long run, i.e. over several centuries or millenia, do not have, never have had, and never will have any weight in shaping policy.

III

NUTRITION

The three most frequently used parameters for comparing nutritional standards are the daily per capita intakes of

(i) food energy (kilocalories)
(ii) total protein (grammes)
(iii) protein of animal origin (grammes)

The ratio of (iii) to (ii) is a rough indicator of the quality of the protein intake, a high proportion of animal protein being correlated with a high intake of the amino acids essential for the growth and replacement of human tissue.

THE ENERGY INTAKE

An intake of 2600 kcal would be considered by most nutritionists to be fairly close to the mean daily energy requirement of a population. The mean requirement varies to some extent with age-distribution, dominant type of physical activity, climatic conditions and the mean body-weights considered as "normal", but these differences are not significant in the present context.

That 2600 kcal is close to the minimum acceptable energy intake can be shown by considering the situation in Britain in 1946. At that time per capita food availability was 2850 kcal, and actual consumption approximately 2650 kcal. As a result of the war, the intakes of the non-farm population of France, Belgium, Holland and Norway were in the range 1800-2300 kcal, of Greece, Yugoslavia, Czechoslovakia and Italy 1400-1800 kcal, and of Germany and Austria 1200-1400 kcal. The energy intake stipulated at the time by UNRRA as necessary for full health and working efficiency was 2650 kcal. Yet the British government, acting on the advice of nutritional experts, declined to take any action to aid the Continental countries which would involve a reduction in the British food intake. Whether this policy was justified or not will not be discussed here. The point to be noted is that when it comes to the crunch 2600 kcal is considered to be a minimum, and nothing short of coercion will get a reduction below this level accepted.

THE PROTEIN INTAKE

A report by the Department of Health and Social Security (UK.) in 1969 recommended intakes of protein linked to intakes of dietary energy. A protein intake equivalent to at least 10% of the energy intake was recommended. As the energy content of protein is 4 kcalg. this recommendation amounts to 65g protein per day on a total energy intake of 2600 kcal.

The FAO/WHO standards of 1974 recommend a minimum protein intake for adults corresponding to 5% of the dietary energy intake.

The British recommendation is a "common sense" figure, slightly below the 11.3% provided by an average British diet. The FAO/WHO figure is based on data from nitrogen balance experiments, and is "far below the level of intake freely chosen. There may be long-term needs for protein in the synthesis of substances turning over slowly, which are not detected in balance studies of short duration. For this reason and also because of the difficulty of obtaining good diets using low protein food, the use of the FAO/WHO figures appears unsound for practical recommendations. A food supply in which 10% of the energy is provided by protein is a safer target for food administrators in all countries." (S. Davidson, R. Passmore, J. F. Brock, A. S. Truswell, "Human Nutrition and Dietetics", Edinburgh, 6th ed., 1975).

THE INTAKE OF PROTEIN OF ANIMAL ORIGIN

If a daily food energy intake of 2600 kcal were obtained entirely by consuming vegetable foods, the protein intake, on average, would be

55g; if it were obtained entirely by consuming animal foods, the protein intake would be over 140g. As 55g of vegetable protein is certainly insufficient, and 140g of animal protein is greatly in excess of requirements, the protein problem is that of deciding on the minimum proportion of animal food in the diet for satisfactory nutrition.

Figure 5 shows total protein and animal protein availability in a number of countries in 1974; Fig. 6 shows food energy and total protein

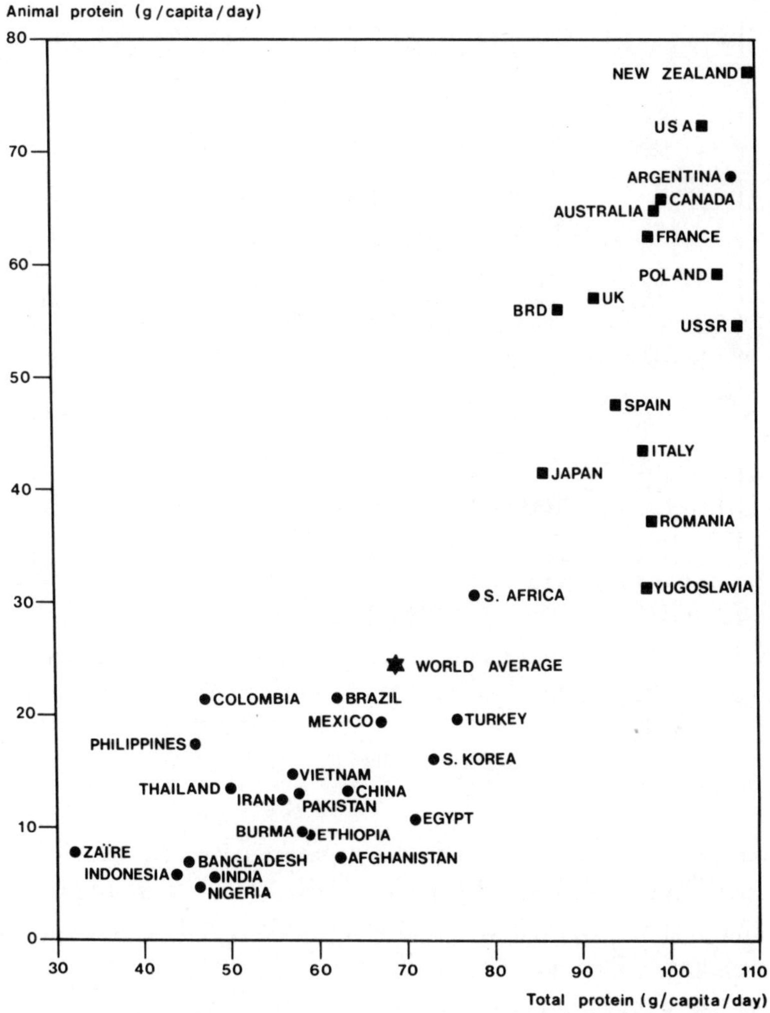

Fig. 5. Total protein and animal protein availability, 1974. (Source: FAO Production Yearbook, 1976).

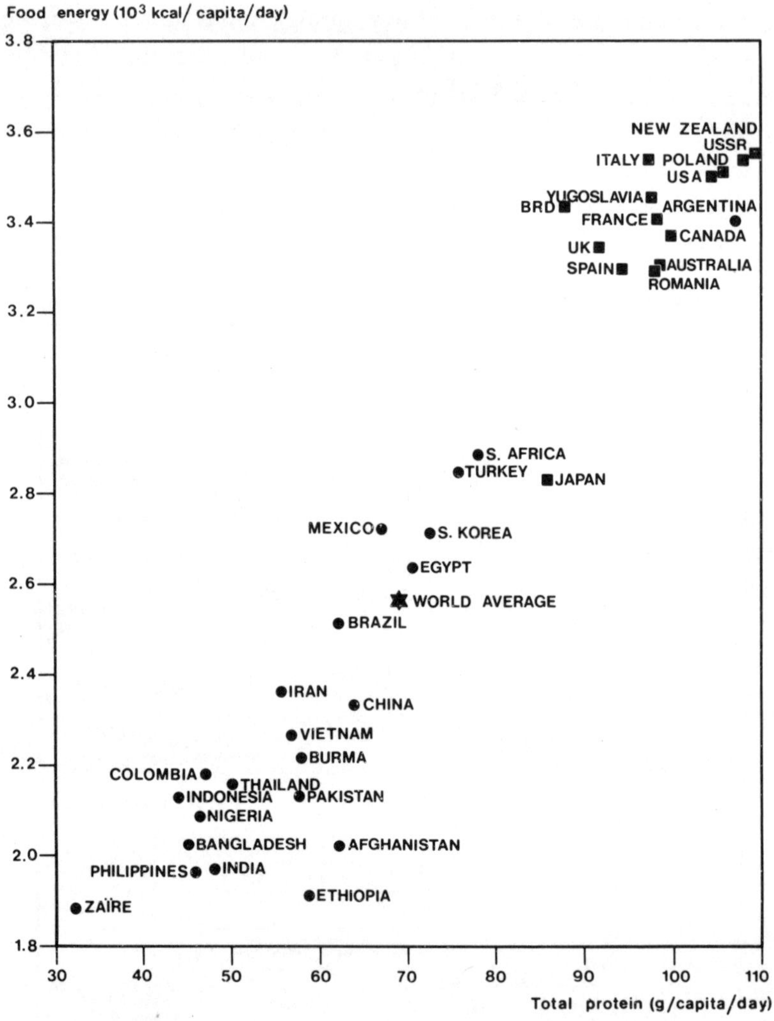

Fig. 6. Protein and food energy availability, 1974. (source: FAO Production Yearbook, 1976).

availability in the same year. The diagrams are based on FAO data, and indicate the amounts available to consumers. Actual consumption is less than the availability, as losses and waste after purchase are not taken into account.* The uncertainty inseparable from estimates based on

*Professor D. S. McLaren suggests that total waste (spoilage, plate waste, food for domestic pets) may amount to 500 kcal/capita/day in affluent societies. ("Nutrition and its disorders", Edinburgh, 1972).

national food balance sheets can be gauged from the data given in Appendix G, in which FAO and OECD estimates of food energy, total protein and animal protein availability in the developed market economies are compared.

There are three main grounds for holding that a substantial proportion of the food intake should be of animal origin—nutritional considerations, security considerations, and consumer preferences. These will be discussed in turn.

NUTRITIONAL CONSIDERATIONS

Proteins consist of combinations of over twenty amino acids, of which eight—isoleucine, leucine, lysine, methionine, phenylalanine, threonine, tryptophan and valine—are essential for the human diet. Two others—cystine and tyrosine—can reduce the requirement for methionine and phenylalanine respectively, and therefore have an intermediate status. The relative proportions of the essential amino acids in the protein of the hen's egg are considered to be closest to human requirements, and the amino acid balance of any protein can be compared to that of egg by computing a "protein score." Protein score is thus a purely qualitative index, not to be confused with protein content. Essential amino acid contents and protein scores of a number of foods are given in Table 6. Protein scores and protein contents of the principal foods are shown in Fig. 7.

Although a diet in which the protein is derived mainly from a single cereal or from starchy roots is likely to be deficient in at least one of the essential amino acids, one protein can supplement another, so that a suitable mixture of vegetable proteins, ingested simultaneously, can provide a sufficiency of the essential amino acids. A satisfactory "mix" of vegetable foods must, however, contain a high proportion of pulses to balance the lysine deficiency of cereal protein. A mixture of equal amounts of wheat and bean proteins, for example, results in a well-balanced protein, the beans compensating the lysine deficiency of the wheat, and the wheat the methionine and cystine deficiency of the beans. The trouble is that pulses are much less digestible than cereals and can be consumed only in relatively small quantities. Maintaining a high level of protein utilization on a diet very low in animal protein is therefore difficult.

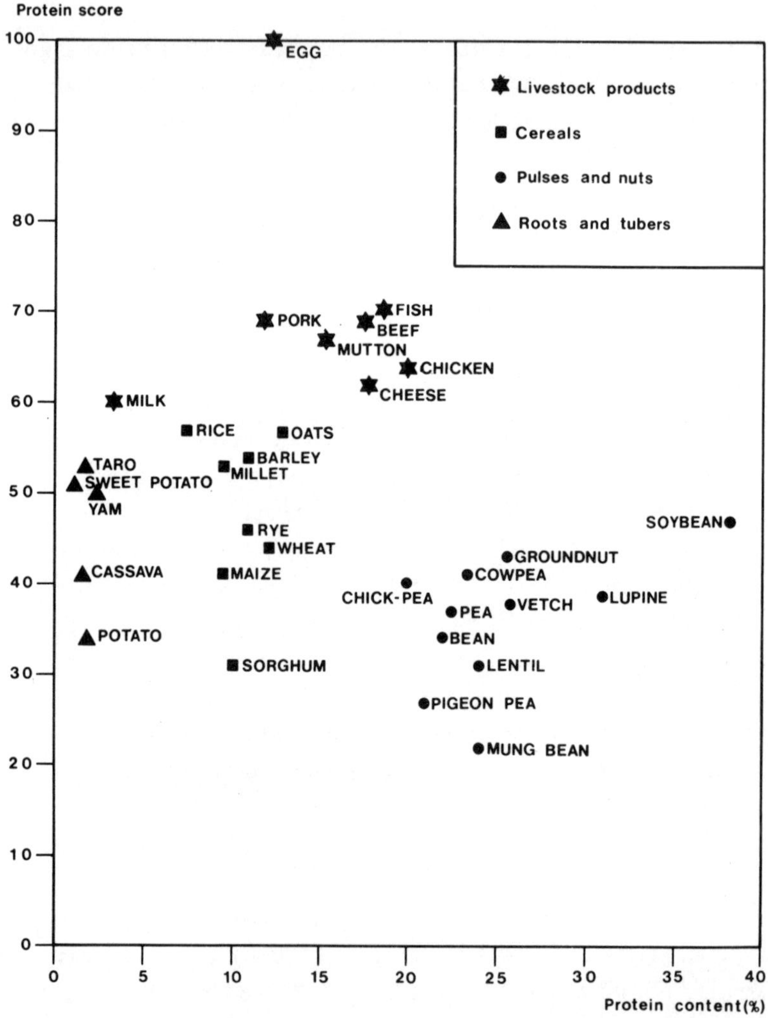

Fig. 7. Protein content and protein score. (Source: FAO Nutritional Study No. 24).

Protein is of particular importance for children, who need it not only to replace tissue, but also for growth, and therefore require protein in a more concentrated form. Foods of animal origin not only contain higher quality protein than vegetable foods, but have protein contents higher than those of most vegetable foods, as can be seen from Fig. 7. As animal protein is several times as expensive as vegetable protein, there is little chance of children getting a specified proportion of animal protein in their diets unless a similar proportion is available to the population at large.

Table 6.

Food	Protein content (g/100 g)	Nitrogen content (g/100/g)	Essential amino acid content (mg/g nitrogen)								Protein score
			Iso-leucine	Leucine	Lysine	Methio-nine + Cystine	Phenyl-alanine + Tyro-sine	Threo-nine	Trypto-phan	Valine	
Hen's egg	12.4	1.98	393	551	436	362	618	320	93	428	100
Beef	17.7	2.83	301	507	556	249	500	287	70	313	69
Fish	18.8	3.01	299	480	569	253	474	286	70	382	70
Cow's milk	3.5	0.55	295	596	487	208	633	278	88	362	60
Rice	7.5	1.26	238	524	237	212	540	244	78	344	57
Wheat	12.2	2.09	204	417	179	253	469	183	68	276	44
Maize	9.5	1.52	230	783	167	217	544	225	44	303	41
Barley	11.0	1.88	224	417	216	246	525	207	96	315	54
Soybean	38.0	6.65	284	486	399	162	505	241	80	300	47
Groundnut	25.6	4.69	211	400	221	150	555	163	65	261	43
Potato	2.0	0.32	236	377	299	118	422	235	103	292	34
Sweet potato	1.3	0.21	230	340	214	175	387	236	107	283	51

Source: FAO Nutritional Study No. 24, "Amino Acid Content of Food and Biological Data on Proteins", Rome 1970.

Another nutritional factor which makes an intake of animal food desirable is vitamin B_{12}, cobalamin. This is the only vitamin not found in any plants except when they are contaminated by microorganisms. Ingestion of such contaminated plants probably provides the only source of vitamin B_{12} for strict vegetarians (vegans). The recommended daily intake of the vitamin for adults is 3 μg according to the U.S. Recommended Dietary Allowances and 2 μg according to the FAO/WHO standards. As milk contains 0.5 μg per 100g, and eggs 1.5 μg per 100g, a daily intake of 400g milk and one egg will supply 3 μg. This involves the ingestion of over 20g animal protein. Animal organs such as liver and kidneys are the richest sources of vitamin B_{12}, but these can form only a small part of the livestock product supply. The daily intake in USA in 1975 was 9.5 μg, of which 70% was supplied by meat and fish, 20% by dairy products, 8% by eggs and 2% by fortified cereals.

SECURITY CONSIDERATIONS

Bad harvests due to unfavourable weather or the appearance of new plant diseases and pests will always occur from time to time. Two or three bad harvests in succession will always be a possibility to be reckoned with. As the production of one kilocalorie of food energy in the form of livestock products involves the consumption of at least seven or eight kilocalories of feedstuff energy, a high level of animal protein intake requires a high per capita crop production. The higher the level of per capita crop production, the higher the margin of safety against crop failure. When production is sufficiently high, a succession of bad harvests would involve nothing worse than slaughtering livestock and living on a less varied diet until feedstuff availability returned to its previous level. When the normal level of production is little above the food and seed requirement, as in most LDCs, a bad harvest will result in famine if an immediate increase of food imports is not possible. Strictly speaking, it is crop consumption rather than production which is correlated with animal protein intake, but consumption is very little in excess of production in almost all LDCs, only countries having a high per capita GNP being able to afford a high per capita grain import.

It is true that a degree of security against plant diseases, pests and the vageries of climate could be achieved by building up grain stocks, but even if an adequate reserve could be accumulated it would be difficult to maintain permanently, as its effect would be to depress grain prices and thus discourage agricultural productivity. Economic and social factors considered, it would appear that livestock is a much better form of energy storage than grain in silos. A certain grain reserve is of course an excellent thing, but as a supplement to livestock, not as a substitute for it.

CONSUMER PREFERENCES

The plain fact is that meat, fish, and dairy products are better eating than cereals, roots, and pulses. The almost universal preference for livestock products is shown by the fact that per capita consumption of animal protein always rises with rising per capita GNP. Table 7 shows the coefficients of income elasticity of demand (the percentage change in demand caused by a 1% increase in income) for cereals, calories, and animal protein in a number of regions.

Investigations of food availability among various income groups within a country show a correlation between animal protein intake and family income. The data given in Table 8 apply to Sri Lanka in 1969-70.

Table 7.

Area	Cereals	Calories	Animal protein
North America	−0.5	−0.03	0.23
Oceania	−0.5	−0.06	0.14
European Economic Community	−0.3	0.10	0.57
Mediterranean Countries	−0.3	0.18	0.90
United Kingdom	−0.4	0.01	0.28
Japan	−0.17	0.20	0.94
Argentina, Uruguay	−0.3	0.04	0.22
Latin America (excl. Argentina and Uruguay)	0.14	0.30	0.60
Near East and Africa (excl. South Africa)	0.2	0.40	1.16
Asia and Far East (excl. Japan)	0.5	0.62	1.49

Source: FAO; Agricultural Commodities—Projections for 1970, FAO Commodity Review Special Supplement (Rome, 1962).

Table 8.

Household income	Food availability per capita per day		
	Energy (kcal)	Protein (g)	Animal protein (g)
Incomes<200 rupees/month	2100	44	7.5
Mean of all income groups	2250	54	10
Incomes >1000 rupees/month	2650	65	20

Source: T. T. Poleman, "Income and Food Consumption: Report to the Government of Sri Lanka", FAO/UNDP, Rome, 1973.

The desirability of animal food from the consumer point of view is partly due to the much higher fat content, food without lipid being very insipid. S. Davidson *et al.* (in "Human Nutrition and Dietetics") consider that good traditional cooking, Eastern as well as Western, requires a fat consumption sufficient to provide at least 20% of the dietary energy. If the latter is taken as 2600 kcal, fat should thus contribute at least 520 kcal, which corresponds to 58 g fat.

Mean world fat availability in 1974 was 61.9 g per capita, of which 34.5 g was of animal origin. The dietary energy obtained from foods of animal and vegetable origin was 447 and 2121 kcal per capita respectively, from which it follows that the mean fat content of animal products is six times that of vegetable products. As the mean intake of animal protein in 1974 was 24.4 g, the latter is roughly the amount needed to bring the total fat intake up to 58 g. It should be noted that this fat requirement is based solely on culinary considerations, a daily fat intake of 15 to 25 g being considered sufficient to provide an adequate supply of fat-soluble vitamins (A, D, E and K) and essential fatty acids (linoleic, linolenic and arachidonic). Total fat intakes in 1974 varied from 179 g in Belgium to 13 g in Bangladesh.

Fats consist of a mixture of saturated, monounsaturated and polyunsaturated fatty acids.* A high Intake of saturated fat is considered as predisposing to coronary heart disease, and certain nutritionists have suggested that intakes of fat, and particularly of saturated fat, in most of the developed countries are too high. For example, it has been sugg-

*The essential fatty acids are polyunsaturated.

Table 9.

Product	Total fat (%)	Polyunsaturated fat (%)	Saturated fat (%)	Saturated fat as percentage of total fat
Soybean oil	100	62	15	15
Sunflower seed oil	100	67	11	11
Cottonseed oil	100	53	29	29
Rapeseed oil	100	24	7	7
Coconut oil	100	2	93	93
Groundnut oil	100	22	19	19
Palm oil	100	12	52	52
Fish oil	100	2	20	20
Butterfat	100	5	50	50
Pork	40	4.3	15.2	38
Beef	22	0.7	9.6	44
Mutton	17	0.6	6.8	40
Chicken	14	3	4.8	34
Milk	3.5	0.1	2.0	57
Cheese	26	0.7	16	62
Eggs	11.2	1.7	3.5	31
Wheat flour	1.5	0.9	0.3	20
Rice (unpolished)	2.2	0.9	0.4	18
Maize flour	2.8	1.4	0.3	11
Barley	1.5	0.6	0.3	20
Oatmeal	5.5	2.1	0.9	16
Millet	3.9	1.6	1.2	31

Sources: P. Helms, "Fødevaretabeller", Copenhagen, 1975. H. H. Hvene, "Smør roligt lidt tyndere paa," Ingeniøren 13, p. 22, Copenhagen, 31 Mar. 1978.

ested (in "Dietary Goals for the United States," 1977) that the total fat intake should not provide more than 30% of the dietary energy, and that saturated fats should not provide more than 10%. If the dietary energy requirement is taken as 2600 kcal, this recommendation amounts to setting limits to total fat and saturated fat intakes of about 90g and 30g respectively.

Total fat content, saturated fat content, and polyunsaturated fat content of the principal fat sources are given in Table 9.

The limits to fats and saturated fats mentioned above imply a limit to animal protein intake. This can be calculated as follows:

Vegetable fat (VF) contains on average 15% saturated fat
Animal fat (AF) contains on average 45% saturated fat
The animal protein/animal fat ratio is 24.4/34.5 = 0.71

Given that	$(AF) + (VF) = 90g$
and	$0.15\,(VF) + 0.45\,(AF) = 30\,g,$
it follows that	$(AF) = 55g$
and	$(VF) = 35\,g.$

The corresponding intake of animal protein is therefore

$$0.71 \times 55 \qquad \approx 40\text{ g.}$$

Table 10.

	UK	France	Italy	Japan
Animal fat intake (1961–5), (g)	110.1	87.3	40.4	18.4
Animal fat intake (1974), (g)	106.2	104.7	58.6	35.1
Expectation of life at birth (years)	(1971–3)	(1973)	(1970–2)	(1975)
M	68.8	68.9	69.0	72.2
F	75.1	76.5	74.9	77.3
Percentage of deaths from:	(1974)	(1974)	(1974)	(1975)
Hypertensive, ischaemic and other heart diseases (B27–29)	32.6	19.2	26.4	16.4
Cerebro-vascular diseases (B30)	13.7	13.9	13.9	24.8

Sources: FAO Production Yearbook, 1976. Annual Abstract of Statistics, 1976. Annuaire Statistique de la France, 1976. Annuario Statistico Italiano, 1976. Statistical Yearbook of Japan, 1977.

Recommendations for limiting intakes of fat or saturated fat do not seem to rest on firm foundations. There is no correlation between expectation of life and animal fat intake in industrialized countries, or between the proportion of deaths from heart and circulatory diseases and animal fat intake, as shown in Table 10. Even if it were proved beyond a shadow of doubt that saturated fat was a predisposing factor in heart disease, there would be no reason to reduce saturated fat consumption unless it could *also* be shown that the mean age of death from heart disease was significantly lower than the mean age of death from all other causes. Even then it would not necessarily follow that saturated fat consumption should be reduced – death from a heart attack may well be preferable to most of the alternatives.

It is reasonable to conclude that the advantages of a high intake of animal protein far outweigh the alleged disadvantages of the concomitant animal fat intake.

DIETARY PATTERNS

The most significant difference between the dietary patterns in DCs and LDCs is the proportion of food of animal origin (see Figs 20 and 21 in Appendix G). Assuming that the dietary energy requirement is 2600 kcal and that the protein-energy ratios for animal and vegetable foods are 5.5g/100 kcal and 2.1g/100 kcal respectively*(Figs. 8 and 9), four dietary patterns – A, B, C and D in Table 11 – can be compared. Pattern A is characterized by animal and vegetable foods contributing equally to the energy intake, pattern B by animal and vegetable proteins contributing equally to the protein intake. In pattern C the supply of animal protein is just sufficient to provide a total protein intake corresponding to 10% of the energy intake. Pattern D is the limiting case of a strict vegetarian diet.

Dietary patterns between A and B will be called Class I diets, between B and C Class II diets, and between C and D Class III diets. All

Table 11.

	Energy intake		Protein intake		
Dietary pattern	Vegetable (kcal)	Animal (kcal)	Vegetable (g)	Animal (g)	Total (g)
A	1300	1300	28	71	99
B	1870	730	40	40	80
C	2313	287	49	16	65
D	2600	–	55	–	55

*The low protein–energy ratios for vegetable foods in USA, UK, BRD and France (Fig. 9) are due to the high sugar consumptions in these countries.

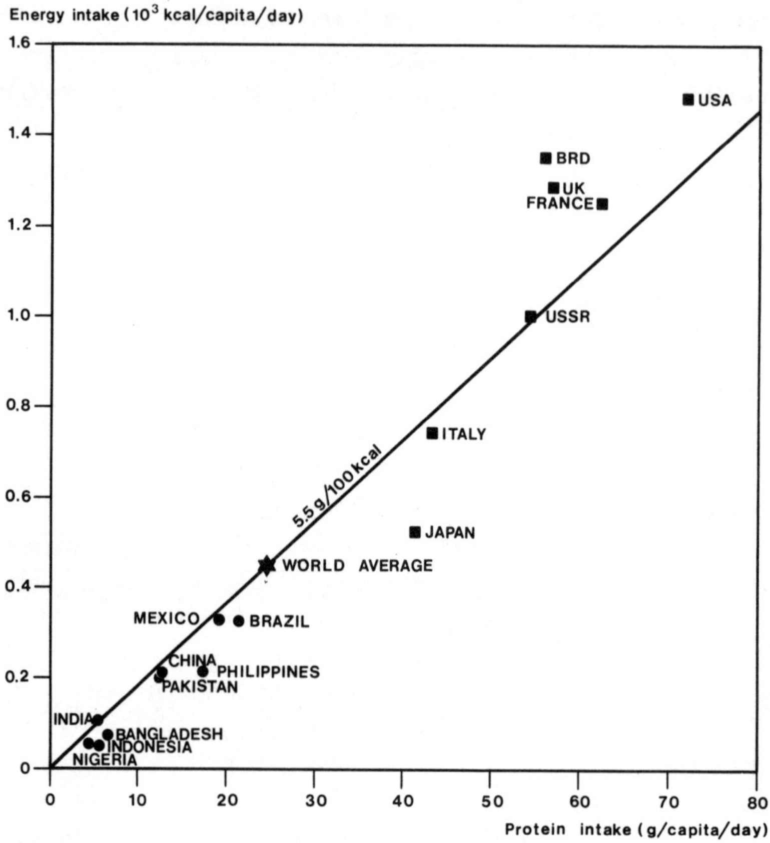

Fig. 8. Protein–energy ratio: Animal food, 1974. (Source: FAO Production Yearbook, 1976.)

DCs excepting Japan and the countries of Southern Europe are in Class I; Latin America (excepting Argentina and Uruguay) is in Class II; Africa (excepting South Africa), South Asia and East Asia are in Class III.

It is clear that Class I diets are satisfactory not only nutritionally, but also from the points of view of security and satisfaction of consumer preferences. Class II diets, though they can be nutritionally adequate, are not very satisfactory gastronomically; if the national average is in the lower range of this category, the nutritional standard of the lowest income groups may well be inadequate. This is because animal protein

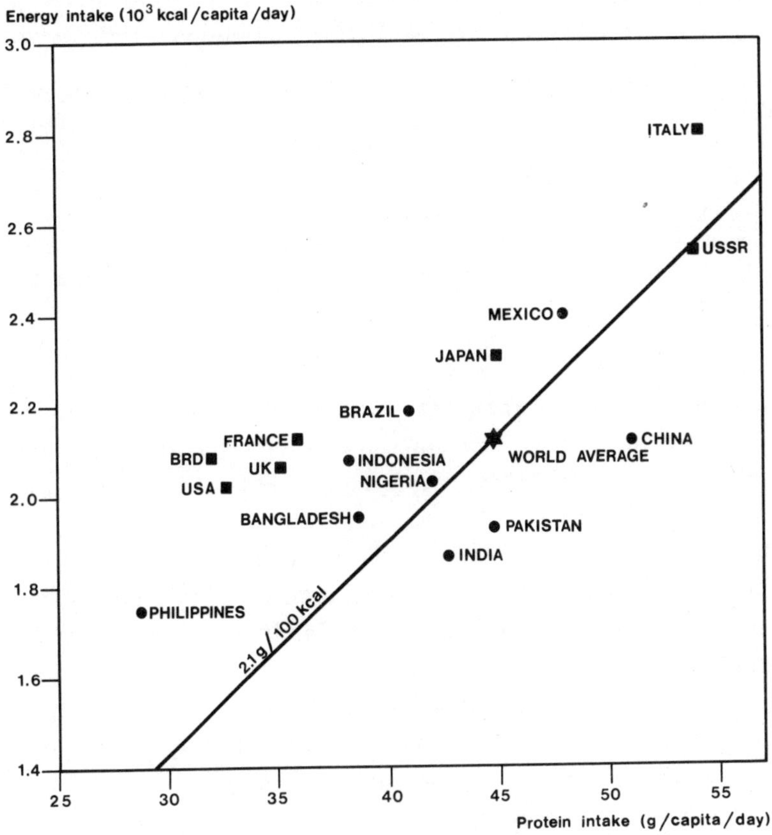

Fig. 9. Protein–energy ratio: Vegetable food, 1974. (Source: FAO Production Yearbook, 1976.)

is and always will be four or five times as expensive as vegetable protein, and the average income of the poorest fifth of the population never exceeds half, and is normally about a quarter, of the national average per capita income. Class III diets are nutritionally inadequate as well as monotonous, and imply a per capita crop production of 5000 kcal/day or less, and therefore cannot be considered satisfactory from the point of view of security.

As it is possible for individuals to obtain adequate nutrition with a relatively low intake of animal protein, national dietary recommendations do not specify any intake of animal protein as being necessary or desirable. A number of nutritionists have, however, done so. Dr. L. R. Wolberg, in "The Psychology of Eating" (London, 1937), writes:

> In constructing the diet so that it will contain ample proteins, it is not at all essential to weigh or to measure food quantities. It is essential to consume one or two generous servings of meat, fish, or cheese daily, one or more eggs and at least a pint of milk daily. These foods will supply the body with from 45 to 70 grams of protein of good biologic quality. The remainder of the menu will contain other foods whose protein, though not of the best quality, will be adequately fortified, and, with the milk, meat and egg protein, will bring the daily protein level to approximately 100 grams.

Professor A. A. Pokrovsky, of the Institute of Nutrition of the Soviet Academy of Sciences, has proposed 40g of animal protein as a minimum for adequate nutrition. Pokrovsky's recommendations are cited and accepted by M. Mesarovic and E. Pestel in "Mankind at the Turning Point," (London, 1975).

Dr. D. W. Kent-Jones, in his article "Cereals" in the 1974 edition of the Encyclopaedia Britannica (Macropaedia), states that ideally half the protein intake should be of animal origin. In practice this means an animal protein intake of 40g.

Sir John Boyd Orr, the first Director-General of FAO, considered 30g of animal protein per day as a convenient line of demarcation between adequate and inadequate standards of nutrition ("The White Man's Dilemma," London, 1953).

PROTEIN OF MARINE ORIGIN

The biological quality of protein of marine origin is, on average, as high as that of livestock product protein. The proportion of the world supply of protein for human consumption which is derived from the sea is, however, small, and is certain to remain so.

The world catch of fish, crustaceans and molluscs in 1976 was 73.5 million tons live weight. The mode of utilization of this catch is summarized in Table 12.

Assuming that the inedible parts (heads, tails, bones, entrails, shells etc) of marine products utilized for human consumption amount to one third of the live weight, and that the mean protein content of the edible parts is 18%, the mean world per capita consumption of protein of marine origin in 1976 was

$$0.67 \times 50.9 \times 10^{12} \times 0.18 / 4 \times 10^{9} \times 365 = 4\text{g/day}.$$

The daily consumption of marine protein in most countries is in the range 2 to 6g. The highest national average is that of Japan – 16g.

The UN estimate for the maximum sustainable world fish catch is 118 million tons. Opinions vary as to the potential catch of comestible marine products for which there is little demand at present, such as krill and octopus, but it would be generally agreed that it amounts to at least 50 million tons, and may be two to three times this figure. It is therefore reasonable to assume that the mean global per capita consumption of marine protein can be maintained at its present level at least until world population reaches 10-12 billions.

Fish culture may make a significant contribution to marine food resources if production can be based on vegetable feeds. The principal feed at the present time is trash fish, about 8kg of which is required to

Table 12.

Mode of utilization	Live weight (10^6 t)
World catch	73.5
For human consumption	50.9
Marketed fresh	19.5
Frozen	13.3
Cured	8.3
Canned	9.8
For all other purposes	22.6
Reduced*	21.6
Miscellaneous purposes	1.0

*Only whole fish destined for the manufacture of oils and meals is included. Raw material for reduction derived from fish primarily destined for marketing fresh, freezing, curing, canning and miscellaneous purposes is excluded; such waste quantities are included under the other disposition channels.

Source: FAO Yearbook of Fishery Statistics, 1976.

produce 1kg of fish for human consumption. (A. H. Weatherley, B. M. G. Cogger, "Fish culture: Problems and Prospects," Science 197, 1977, p. 427-430).

TECHNOLOGICAL "SOLUTIONS"

Raising per capita intakes of animal protein in the LDCs would involve increasing crop production at a rate substantially higher than the rate of population growth, and maintaining this agricultural growth rate over a number of years. As this is practically impossible to achieve as long as population growth rates in the LDCs remain at present levels, other methods of improving nutritional standards in these countries have been proposed.

(i) Synthetic amino acids

Improvement of the amino acid balance of a vegetable diet by means of synthetic amino acids has so far been limited to animal feeds. Marketing synthetic amino acids for human consumption would present many problems, but even if these could be overcome, it can hardly be maintained that fortifying the diets of several billion people for an indefinite period by means of tablets would be a satisfactory solution of the problem of protein supply. The daily consumption of tablets could be avoided to a certain extent by fortifying flour with synthetic amino acids, particularly lysine. This has been done on a limited scale in India, but the efficacy of the method has been questioned,* and even its advocates do not suggest that it can render a higher consumption of animal protein unnecessary. Fortified flour can in any case reach and benefit only those who purchase a high proportion of their food, i.e. urban residents.

(ii) Improved grain varieties

New varieties of cereals are being developed with the fourfold object of improving the amino acid balance of the protein, increasing the protein

*"Relatively extensive field trials in Tunisia, where wheat products were supplemented with lysine... and in Thailand, where rice was supplemented with lysine and threonine... have led to essentially negative results."
(D. M. Hegsted, "Protein-Calorie Malnutrition," American Scientist 66, Jan. 1978, p. 61-65).

content of the grain, increasing the yield potential and the resistance of the plant to pests and diseases. One such variety is Opaque – 2 maize, which has a protein score of 48 and a protein content of 11.6% compared with 41 and 9.5% for traditional maize; these advantages, however, are obtained at the price of lower yield and greater susceptibility to insects and disease. The new cereals also face problems of consumer acceptability; it is reported from India that the traditional flinty, hard grain is preferred to the soft-kernel, protein-rich, high-lysine variety ("Time," 11 Nov. 1974); freshly cooked rice of low protein content is considered by most rice-eaters to be more tender, cohesive, and of better flavour than high-protein rice (D. H. Grist, "Rice," London, 1975).

(iii) Protein from petroleum

This proposal for augmenting protein supply involves the use of single-celled micro-organisms to convert petroleum into edible protein. Apart from the technical problems which must be overcome before single-cell protein can be safely consumed by humans, the main limitations to this method of food production is that world petroleum production is expected to peak by the end of the present century, and a subsequent decline is inevitable. Single-cell protein cannot make a significant contribution even to feedstuff supply for more than a few decades.

(iv) Protein from crop residues

Another proposed method of synthetic food (or feed) production is the conversion of inedible harvest residues (straw, corncobs etc) and by-products (bagasse etc) to molasses, and the conversion of the molasses to a protein-rich material by means of yeast fermentation. Though there will always be an abundant supply of the raw material, it is uncertain whether this proposal will ever be economic. There is also the problem of the acceptability of the product.

The position as far as synthetic foods are concerned has been admirably summed up by the agronomist Maurice Guernier, in his article "L'équilibre fragile" (Dossiers Jeune Afrique 2, 1975):

> For thirty years there has been talk of algae, of chlorella, and of protein derived from petroleum products. I say simply this: show me a man who has eaten

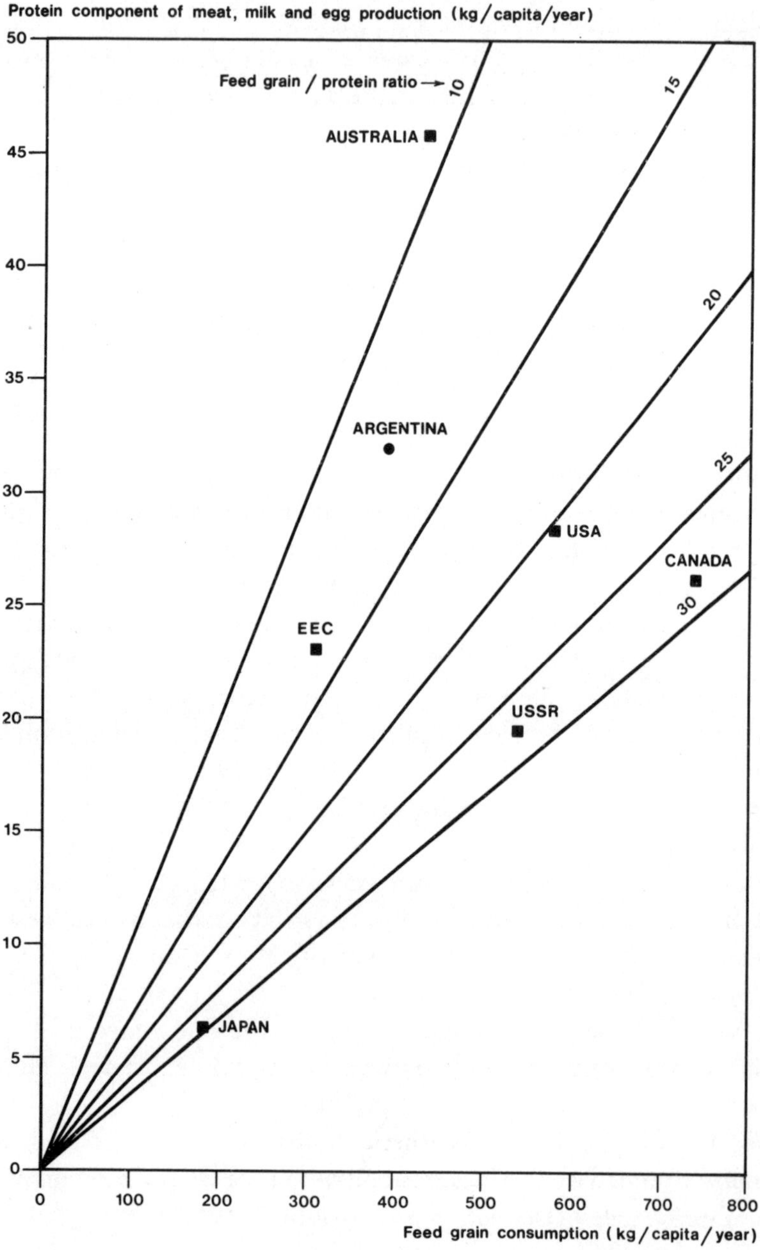

Fig. 10. Conversion of feed grain to animal protein. (Based on calculations in Appendix A).

two meals composed of synthetic products! I say two meals, for if he has eaten one, he will refuse the second! Of course all methods must be tried, but with the primary object of substituting synthetic products for the fish-meal, groundnut meal, and soybeans now consumed by animals. (Transl. BG)

THE AGRICULTURAL POTENTIAL

The variegated picture presented by world agriculture – 15 million km^2 of arable land and 30 million km^2 of permanent pasture, the former cultivated by methods varying enormously in degree of capitalization and technological development, and both consisting of a wide range of soil types subjected to a great variety of climatic conditions – make long-term forecasts of agricultural production very risky. It is nevertheless possible to make agricultural projections with a reasonable degree of confidence, it being understood that a projection is not a forecast, but an estimate of what could be achieved if certain specified conditions were fulfilled. Even so, the problem is highly complex. It can, however, be rendered tractable by using cereals as the primary indicator of agricultural production. This simplification is justified by the fact that the area planted to cereals (7.5 million km^2) constitutes almost three quarters of the area cropped annually. As far as pasture is concerned, it should be borne in mind that although the area of permanent grassland is double the arable area, its contribution to world food supply is relatively low, partly because the mean grass yield is only a quarter of the mean crop yield, and partly because grass can be utilized only by being processed through ruminants, and the efficiency of conversion of grass to meat by cattle and sheep on rangeland is very low.

Figure 10 shows the relation between per capita consumption of feed grain and per capita production of meat, milk and egg protein in a number of countries. Australia and Argentina have low grain-protein ratios because the agricultural area in these countries is many times the cereal area. The EEC has a low ratio because of the high productivity of its grasslands. For further details on the input-output ratio in livestock production the reader is referred to Appendix A.

IV

CROP YIELDS

GRAIN YIELD

The biological limit to grain yield is the product of the mean intensity of photosynthetically active radiation incident on the ground during the grain-filling period, the length of the grain-filling period, and the limiting efficiency of photosynthesis in cereal plants:

(i) Incident radiation

The intensity of solar radiation at the outer limit of the Earth's atmosphere is 1353 W/m^2. Only 47% of this radiation reaches the Earth's surface, the remainder being reflected into space or absorbed by the atmosphere. The mean intensity of incident radiation in the temperate zones during the summer, and in the tropics all year round, is very nearly

$$1353 \times 0.47/\pi = 202\ W/m^2 = 418\ cal/cm^2/day.$$

(The factor π arises because the intercepted radiation is distributed round the tropical zone as the Earth rotates. The mean incident radiation on the entire surface of the Earth is $(1353 \times 0.47/4) = 159\ W/m^2$. as the surface of the globe is four times the area it intercepts.)

The mean radiation on the world's cropland during the grain filling period will be taken as 400 $cal/cm^2/day$. The proportion of this radiation which is photosynthetically active, i.e. in the wavelength range 400 – 700 nm, is 45%.

(ii) The grain filling period

The grain-filling period for cereals in N.W. Europe is approx. 50 days, for maize in the U.S. Corn Belt and rice in Japan approx. 40 days. There is some evidence that the product of grain-filling period and mean daily incident radiation is 20000 cal/cm^2 for wheat and maize, and 16000 cal/cm^2 for rice. The grain-filling period at 400 cal/cm^2/day will be taken as 40 days for rice and 50 days for other cereals.*

(iii) The efficiency of photosynthesis

Photosynthesis is essentially a process whereby eight photons of white light supply the energy necessary to liberate the carbon atom from a molecule of carbon dioxide, the oxygen being dissipated and the carbon uniting with a molecule of water to form carbohydrate. The efficiency of this process, i.e. the ratio of the energy stored in the carbohydrate to the energy used in reducing the carbon dioxide, is 28.6%. (The calculation is given in Appendix C). This efficiency – known as the quantum efficiency – cannot be attained as an overall efficiency in crop plants because of various losses inseparable from plant physiology. The maximum efficiency of conversion of photosynthetically active radiation attainable in cereal plants, using values for the physiological losses given by Dr. L. T. Evans, is 10% in C_4 plants (maize, sorghum, millet), and 7% in C_3 plants (wheat, rice, barley, oats, rye). Y. Murata and S. Matsushima have estimated the maximum efficiency for rice at 10%. The calculations are given in Appendix C.

THE THEORETICAL YIELD LIMIT

Assuming that the energy content of cereal grain with a moisture content of 10% is 3400 cal/g, the yield which would be obtained if the limiting photosynthetic efficiency were maintained throughout the grain-filling period of a C_4 plant is

$$400 \times 10^4 \times 0.45 \times 50 \times 0.10/3400 \approx 2600 \text{ g/m}^2$$
$$= 26 \text{ t/ha}$$

*The roots, stem and leaves, which have a dry matter content slightly greater than that of the grain, have a growth period of about 100 days. The mean rate of dry matter formation during this period is thus little more than half that during grain-filling. This is chiefly because a full leaf canopy is not developed until 40–50 days after sowing.

The corresponding figure for C_3 plants is approximately 18 t/ha. Rates of energy conversion close to the limit have been achieved for very short periods at experimental stations, but these rates cannot be maintained for anything approaching the entire grain-filling period, as the storage capacity of the plant, determined by morphogenesis early in the growth-period, is limiting in practice.

Record yields of cereals are in the region of 70% of the photosynthetic limit, as can be seen from Table 13.

Table 13.

Cereal	Yield (t/ha)
Maize (15.5% moisture)	22.1
Sorghum	20.0
Wheat	14.1
Rice (rough, 14% moisture)	13.0
Barley	11.4
Oats	10.6

Sources: S. H. Wittwer, "Food Production: Technology and the Resource Base", *Science* **188**, 9 May 1975. L. T. Evans (ed.), "Crop Physiology", Cambridge Univ. Press, 1975. Guiness Book of Records, 1979.

Table 14.

Cereal	Country or Administrative area	Year	Yield (t/ha)	Harvested area (10^3 ha)
Maize	New Zealand	1973	9.04	13
Rice (rough)	Yamagata Prefecture, Japan	1975	7.95	102
Wheat	Netherlands	1974	5.73	130
Sorghum	Iowa	1971	5.46	40
Oats	Netherlands	1974	5.02	33
Barley	Belgium	1974	4.68	150
Rye	Switzerland	1974	4.62	9
Millet	Egypt	1974	3.91	211

Sources: FAO Production Yearbook, 1976: Statistical Yearbook of Japan, 1977; USDA Agricultural Statistics, 1974; Statistics of area, production and yield of crop products, OECD, 1977.

Table 15.

Cereal	Yield (1976) (t/ha)	Harvested area (1976) 10^3 ha)
Maize	2.83	118054
Rice (rough)	2.43	142248
Barley	2.03	93446
Wheat	1.77	235302
Rye	1.68	16431
Oats	1.67	30244
Sorghum	1.18	43929
Millet	0.71	72808
All cereals	1.94	759422

Source: FAO Production Yearbook, 1976.

The highest yields for the eight principal cereals, achieved as averages on harvested areas exceeding 5000 ha, are given in Table 14. The last column of the Table gives the area on which the yield was obtained. Mean world yields and harvested areas for each cereal are given in Table 15.

Professor P. Buringh, H. D. J. van Heemst and G. J. Staring, in "Computation of the Absolute Maximum Food Production of the World" (Wageningen, 1975), assume that year-round cultivation under optimum conditions at 400 cal/cm²/day incident radiation can give a dry matter yield of 80 t/ha per year for C_3 plants. They assume that the roots and stubble account for 25% of the dry weight, and that the grain and straw account for equal shares of the balance. With a moisture content of 10%, the theoretical annual grain yield is thus $80.0 \times 0.375/0.9 =$ 33 t/ha. Year-round cultivation, however, implies three harvests, the sum of the grain-filling periods being about 120 days. The yield corresponding to a 50-day grain-filling period is therefore $33 \times 50/120$ or 14 t/ha. This is in good agreement with the record wheat and paddy yields given in Table 13.

The authors calculate that the theoretical limit for annual grain yield varies from 30 t/ha in the lower Ganges valley and delta (year-round cultivation with a high proportion of the cultivated area under irrigation) to 1.4 t/ha in the Yakutsk area (one short growth period without irrigation). Their estimate for the theoretical limit for world yield is

13.4 t/ha. The method of calculation leads to curious results. For example, the potential annual yield of the lower Rhone valley is given as 23 t/ha, while that of the Netherlands and Belgium is 14 t/ha; yet the latter countries have maintained the highest yields in Europe throughout the past forty years. Again, the potential of Iowa and Illinois, by far the most productive cereal areas of N. America, is given as 12 t/ha, compared to 19 t/ha in Louisiana. Yet again, the potential of Egypt is given as 28 t/ha, even though grain yields in that country have levelled off at 4 t/ha and extension of the irrigated area is considered to be the only means of increasing production. These considerations show how tenuous is the connection between theoretical assumptions and field conditions, between what is possible on an experimental station and the maximum sustainable yield on several million hectares.

In a subsequent report ("An Estimation of World Food Production based on Labour-oriented Agriculture", Wageningen, 1977) Buringh and van Heemst suggest that the practical yield limit, i.e. the limit attainable in commercial agriculture, is about 50% of the theoretical limit computed in their previous report. It will be shown below that this estimate is over-optimistic.

Of the many factors which limit yield in world agriculture, two lend themselves to quantitative treatment – water supply and chemical nitrogen fertilizer.

WATER SUPPLY

The water requirement (or transpiration ratio) for most cereals is 300 litres/kg dry matter. As the grain constitutes almost 50% of the dry matter, the requirement is 600 litres/kg grain. The requirement for rice is higher – about 900 litres.

Rainfall is by far the most important source of water for the world's cropland. Solar radiation evaporates 520 thousand km^3 water annually, resulting in a mean global precipitation of 1020 mm. The mean precipitation on the 117.5 million km^2 of land drained by rivers and streams is 860 mm. The mean precipitation on the world's cropland is slightly higher than this, and will be assumed to be 900 mm. If the entire annual precipitation on cropland were transpired by crop plants, the mean world grain yield (assuming all other conditions to be optimum) would be

$$0.9 \times 10^4/650 = 14 \text{ t/ha}.$$

(The global water requirement for cereals is 650 litres/kg grain as the rice area is 19% of the cereal area). This yield is a theoretical limit, impossible to attain because the bulk of the precipitation is lost through evaporation, percolation and run-off.

The maximum attainable efficiency of utilization of the annual precipitation in rainfed agriculture is difficult to estimate. The problem can best be treated by considering the world cereal area as consisting of two approximately equal areas – the temperate zones and the tropics.

The mean annual precipitation in the temperate zones is approx. 600 mm, about 45% of which falls in the five-month growth period of the crop. The average yield limit for rainfed agriculture in the temperate zones is therefore

$$0.6 \times 10^4 \times 0.45/600 = 4.5 \text{ t/ha}.$$

(The rice area in the temperate zones being very small, the water requirement can be taken as 600 litres/kg grain). In the most favoured regions of the temperate zones (N. W. Europe, U. S. Corn Belt) where the rainfall in the growth period is in the range 300-450 mm, the yield limit is from 5 to 8 t/ha. In the USSR, on the other hand, the limit is well below the temperate zone average. As L. Symons puts it (in "Agricultural Geography", Bell, London, 1966):

"It is .... generally recognised that physical conditions in the USSR are far inferior for agriculture generally to most of the USA or Western Europe. Even in the 10% or so of the USSR which is within the limits set to cultivation by cold, drought and relief, the more humid areas have long winters and the warmer areas are mostly dry. In such circumstances yields can never compare with those in more favourable conditions in Western Europe and the USA under similar systems of cultivation."

In this connection it is worth noting that the target for Soviet grain production in 1980 is 215-220 million tons, or slightly less than the record harvest of 1973. As the Soviet cereal area has stabilized at 1.2 million ha, the target corresponds to a yield of 1.8 t/ha, a little below the world average. Soviet grain statistics, incidentally, should be discounted by 10-15%, as they are based on bunker-weight, which includes a significant amount of extraneous matter and a high moisture content.*

In most of the tropics the mean annual precipitation of about 1200 mm takes the form of heavy downpours concentrated in a season of four or five months. The maximum efficiency of utilization of the annual rainfall is much lower than in the temperate zones, due to the high rates of evaporation and percolation, and can hardly exceed 25%. This gives a yield limit almost equal to that of the temperate zones, an estimate which agrees with those made by the National Institute of Agricultural Research of Mexico, in which the limiting yield of non-irrigated areas in Mexico with an annual precipitation of 800-900 mm is given as 4.7 t/ha for soils classified as "deep," and 3.3 t/ha for soils classified as "shallow." A mean yield of 4 t/ha with an annual precipitation of 850 mm corresponds to an efficiency of water utilization of 28%. The limiting efficiency in regions with higher annual precipitations will certainly be lower, so that a limit of 4 t/ha for the tropics as a whole cannot be far out; it is if anything on the high side, particularly when the large millet area (almost 20% of the cereal area of the tropics) consisting almost entirely of poor soils in semi-arid climates, is taken into account.

Irrigation is the only means of raising the yield limit set by rainfall. World consumption of irrigation water is 1700 km^3/year, corresponding to a mean depth of 750 mm over the irrigated area of 227 million ha. Grain yields on irrigated areas have reached 8 t/ha, but such a yield can only be achieved when water can be supplied in controlled amounts at any time during the growth period of the crop. This situation does not obtain over much of the world's irrigated area, where the supply of irrigation water varies with river flow. In some countries (e.g. Syria) the irrigated area itself fluctuates from year to year. As it is extremely unlikely that the entire world irrigated area will ever be supplied by technologically advanced storage and distribution systems, involving dams, reservoirs, pumping stations, pipelines and sprinklers, the most reasonable estimate for the yield potential of the world irrigated area is 6 t/ha. This coincides with the estimate of the National Institute of Agricultural Research of Mexico for the yield limit of the Mexican irrigated area.

*I am indebted to Prof. D. Gale Johnson for this information.

H. J. Moen and K. J. Beek, in "Literature study on the potential irrigated acreage in the world" (Wageningen, 1974), have estimated the irrigation potential of the major regions. These estimates are tabulated below, together with the arable area of each region and the area at present irrigated.

Table 16.

Region	Irrigated area in 1975 (10^6 ha)	Irrigable area (10^6 ha)	Arable area in 1975 (10^6 ha)	Irrigable area as percentage of arable area
N. America	17.0	28	253	11
Europe	12.7	33	147	22
USSR	14.5	47	232	20
Oceania	1.6	5	47	11
Latin America	12.2	27	141	19
Africa	7.7	20	211	9
Asia	161.1	310	479	65
World	227	470	1506	31

Source for 1975 areas: FAO Production Yearbook, 1976.

It can be concluded that with full utilization of the irrigation potential, the mean yield in Asia could ultimately attain 5 to 6 t/ha, and in other regions 4 to 5 t/ha.

The cost of increasing the world irrigated area to 470 million hectares and improving most of the existing irrigation systems would run to over a trillion dollars at current prices. The assumption that allocations of this magnitude can be made during the next seventy years is highly optimistic.

NITROGEN AVAILABILITY

As protein contains 16% nitrogen, the latter is an essential nutrient for all crops. The average protein content of cereal grain is 10%, that of the straw being about 4%. As the weights of grain and straw are approximately equal, it follows that the total nitrogen requirement is 1.2% of the plant weight or 2.4% of the grain weight.

Precipitation, irrigation water and non-symbiotic micro-organisms provide sufficient nitrogen for a grain yield of 1 to 1.5 t/ha. Higher yields can only be achieved by means of fertilizers. Organic fertilizers (compost, manure) have a very low nitrogen content, and the yield increases obtainable by their use are limited; the higher the yield, the higher the proportion of soil nitrogen which must be supplied by chemical fertilizer. Professor P. Buringh and H. D. J. van Heemst, in "An Estimation of World Food Production Based on Labour-oriented Agriculture" (Agricultural University, Wageningen, 1977) estimate that without the use of chemical fertilizers, the maximum attainable global yield is close to 2 t/ha, regional limits varying from 2.3 t/ha in North America to 1.5 t/ha in Africa.

The yield increase obtainable by means of chemical fertilizer depends on the amount of nitrogen applied per unit area, and on the proportion of the applied nitrogen absorbed by the crop plants. It is generally accepted that the incremental yield-nitrogen ratio for cereals (the additional grain yield obtained by increasing the intensity of nitrogen application by 1 kg/ha) declines rapidly at nitrogen intensities of about 200 kg/ha. This intensity will be assumed to be the practical limit. Intensities of application of chemical nitrogen in 1974 varied from zero in many African countries to the world record of 208 kg/ha in the Netherlands.*

If a nitrogen application of 200 kg/ha were entirely absorbed by the crop plants, the resulting yield would be approximately

$$1 + (0.2/0.024) = 9 \text{ t/ha}.$$

The actual efficiency of utilization is far below this, the bulk of the nitrogen being lost through volatilization, percolation and run-off. The relation between intensity of nitrogen application and grain yield is calculated in Appendix D, and illustrated in Fig. 11. On a world basis the relation is found to be

$$Y = 1171 + 17.8\,(N/A)$$

where Y = grain yield in kg/ha

N/A = intensity of nitrogen application in kg/ha.

The yield corresponding to N/A = 200 is 4730 kg/ha. It is of interest to

*In countries which have very high livestock densities, such as the Netherlands and Denmark, nitrogen from manure amounts to approx. 50 kg/ha per year. The total nitrogen application can thus be as high as 250 kg/ha.

note that the mean world grain yield in 1934-8 (when the mean intensity of nitrogen application on cropland was 4 kg/ha, the greater part of which was used for non-cereal crops) was 1150 kg/ha.

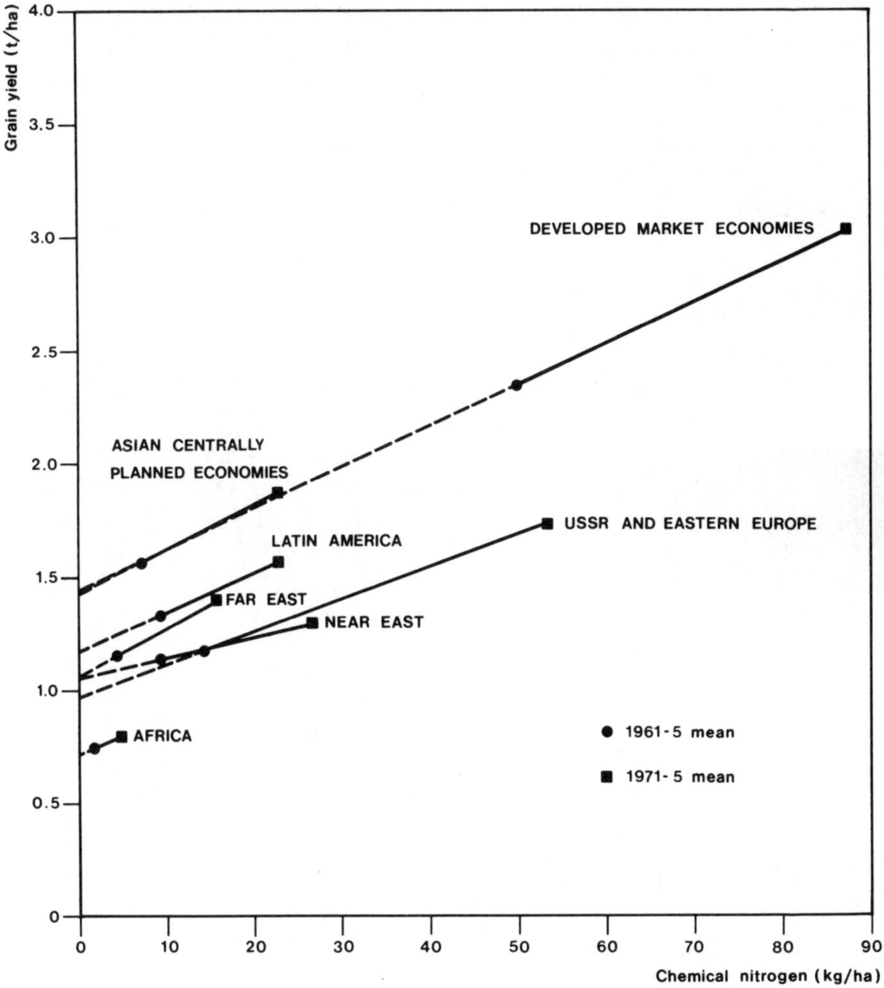

Fig. 11. Grain yield and chemical nitrogen fertilizer. (Source: FAO Production Yearbooks, 1973-5).

RECENT TRENDS IN GRAIN YIELDS

Cereal yields in countries which have achieved great increases in agricultural productivity during the past twenty years have now either levelled off or are increasing very slowly. Yields of wheat, rice and maize over the period 1955-1976 in countries where these crops are cultivated intensively on a substantial area are shown in Figs 12, 13, 14.

The recent stabilization of yields in a number of industrialized countries does not mean that higher yields are unattainable, but it does indicate that higher yields cannot be attained without a considerable increase of inputs (and consequently costs), and that it is unrealistic to imagine that the subsistence farmers of Latin America, Africa and Asia will be able to achieve yields which exceed current

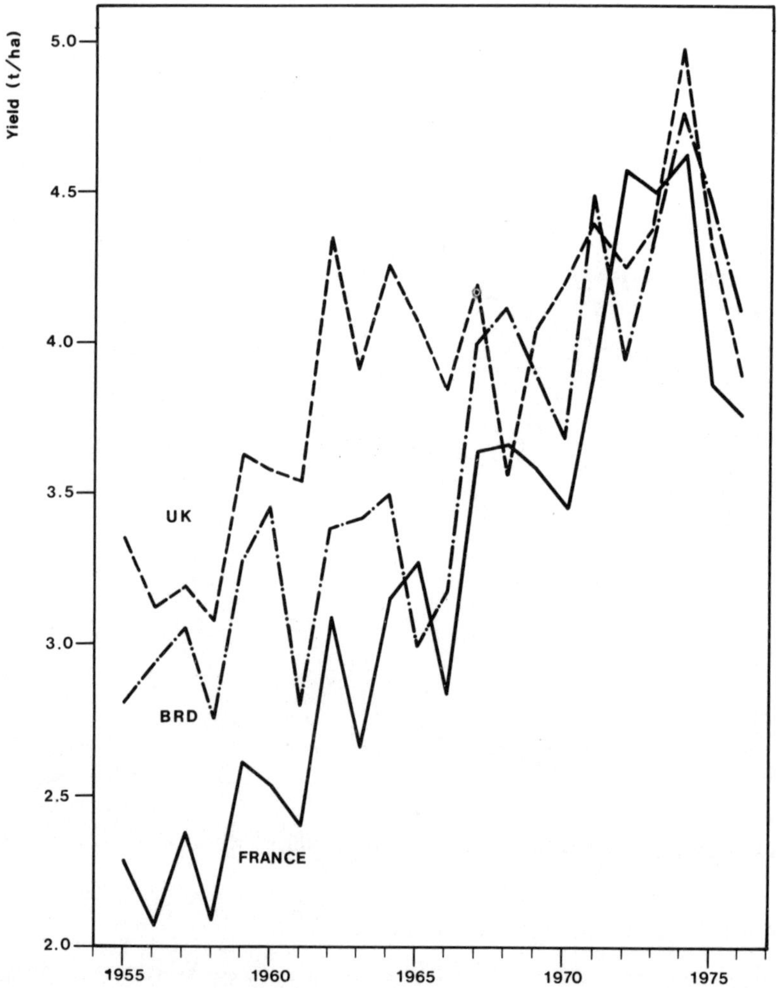

Fig. 12. Wheat yields, 1955–1976. (Source: Statistics of area, production and yield of crop products in OECD member countries, Paris, 1977).

yields in the heavily capitalized agricultures of Western Europe, USA and Japan within the period under consideration, that is, by the year 2050.

YIELDS OF OTHER FIELD CROPS

The most important non-cereal crops are starchy roots, dry legumes, sugarbeet and sugarcane. Estimates for biological yield limits for these

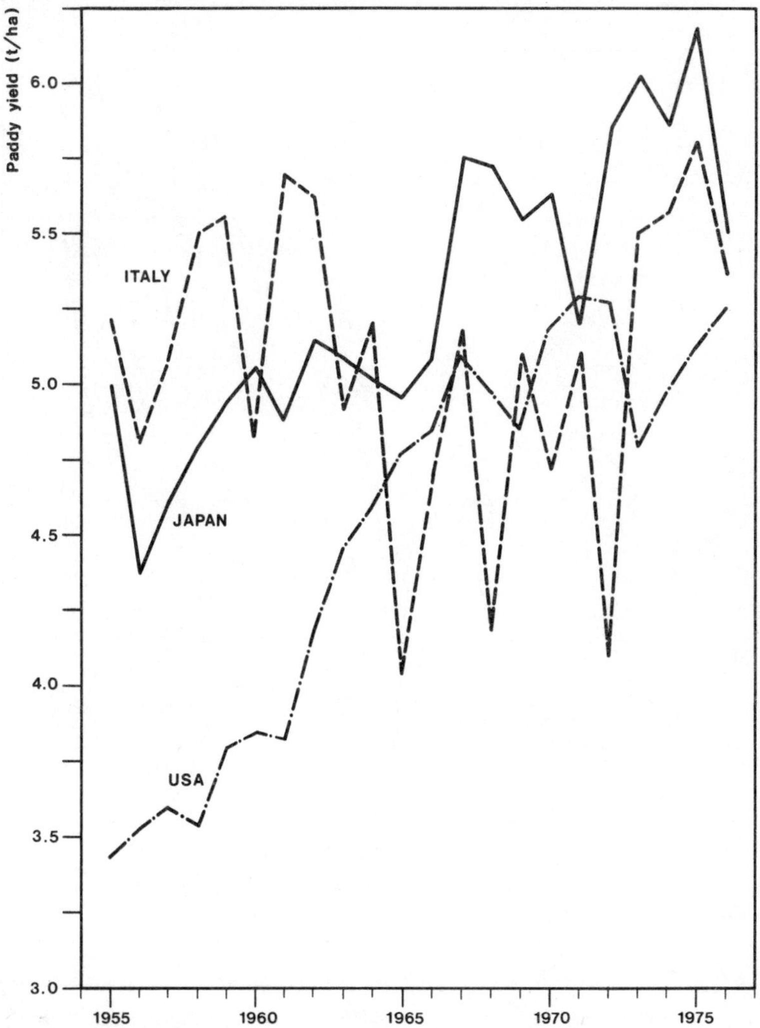

Fig. 13. Rice yields, 1955–1976. (Source: Statistics of area, production and yield of crop products in OECD member countries, Paris, 1977).

crops are subject to a wider margin of uncertainty than the biological limit for cereals, and will not be discussed here. Statistics are given below for world record yields (Table 17), highest dry matter yields achieved as national averages. (Table 18), and mean world yields (Table 19) for potatoes, soybeans and sugar crops.

A common figure is given in Table 19 for sugar from beet and cane, as it is not possible to calculate separate yields at world level. This is

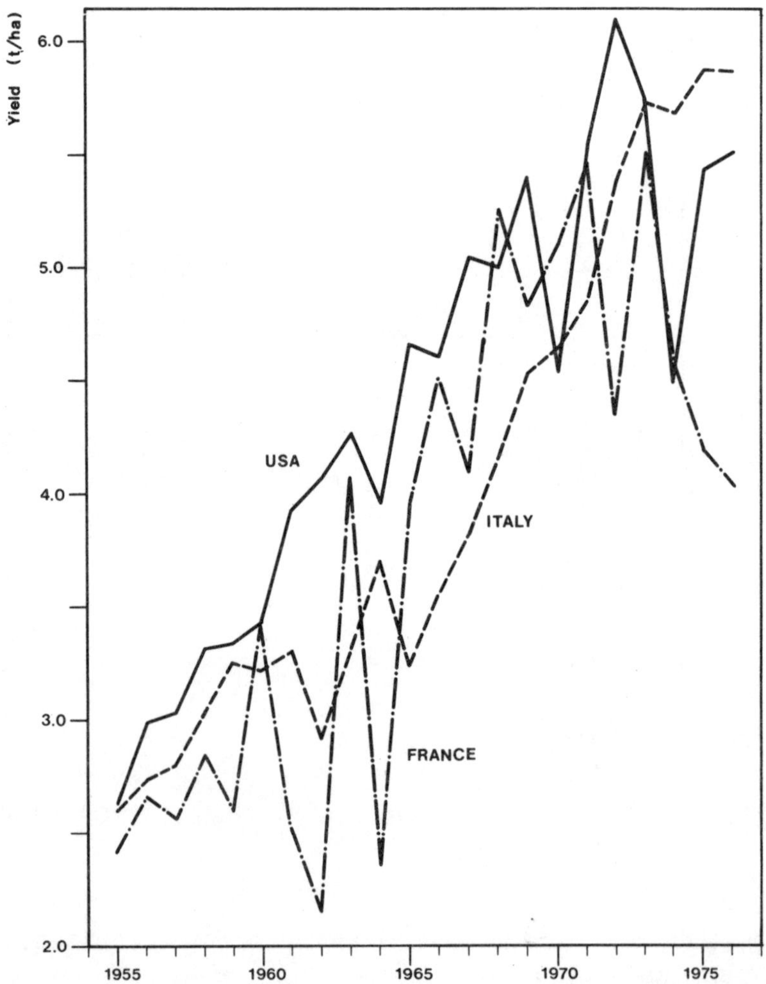

Fig. 14. Maize yields, 1955–1976. (Source: Statistics of area, production and yield of crop products in OECD member countries, Paris, 1977).

Table 17.

Crop	Gross yield (t/ha)	Dry matter (t/ha)	Growth period (days)
Potatoes	95	23	~150
Soybeans	5.56	5	~150
Sugar beet	115	19*	240
Sugarcane	250	~30*	~600

*Sucrose extracted.
Source: L. T. Evans (ed.), "Crop Physiology", Cambridge 1975.

Table 18.

Crop	Country	Year	Gross yield (t/ha)	Dry matter (t/ha)	Harvest area (10^3 ha)
Potatoes	Switzerland	1975	43.22	10.6	21
Soybeans	Canada	1975	2.32	2.1	158
Sugarbeet	Austria	1975	52.33	8.3*	60
Sugarcane	Peru	1974	168.9	19.1*	54

*Sucrose extracted.

Sources: FAO Production Yearbook, 1976; Statistics of area, production and yield of crop products, OECD, 1977.

Table 19.

Crop	Gross yield (1975) (t/ha)	Dry matter (1976) (t/ha)	Harvest area (1976) (10^3 ha)
Potatoes	13.63	3.36	21097
Soybeans	1.38	1.24	44885
Sugarbeet	31.37	4.46*	9376
Sugarcane	54.42	4.46*	12732

*Sucrose extracted.

Source: FAO Production Yearbook, 1976.

because some countries cultivate both beet and cane, and sugar production statistics are given for total production only. Beet cultivation, however, is largely confined to the DCs and cane cultivation to the LDCs, so that sugar yields from beet and cane will be close to the mean sugar yields of the DCs and the LDCs respectively. (The mean DC yield in 1973 was 4.45 t/ha, the LDC yield 4.67 t/ha).

The average dry matter yield of all the major non-cereal field crops (starchy roots and tubers, soybeans, dry legumes, sugar crops, groundnuts, sunflowerseed, linseed, rapeseed and sesameseed) on the 239 million hectares harvested in 1976 was 1.6 t/ha, slightly less than the 1.75 t/ha yield of dry matter obtained from the 759 million hectares planted to cereals.

GRASS YIELD

World production of grass exceeds that of any other category of crops excepting cereals. Most of the grass grows on permanent pasture, a proportion being grown on arable land temporarily not utilized for field crops (rotation grasses).

The following factors must be borne in mind when considering the role of grass in world food production:

(i) Grass can be directly utilized only by ruminants, and only about 60% of the dry matter is digestible by them.

(ii) Much of the grass production of arable areas must be conserved as hay or silage, grass having no storage organ in which dry matter is accumulated. The attendant losses of dry matter can amount to 40%.

(iii) Practically all the world's permanent pasture is neither sown, fertilized nor irrigated. It is therefore impossible to increase yield on these areas.

In spite of the above limitations, the cultivation of grass can result in very high yields on small, intensively fertilized areas. This is because grass maintains a high leaf area index throughout the year, so that a relatively high proportion of the incoming radiation is intercepted, and also because a high proportion of the shoot system can be harvested during the course of the year.

The highest recorded annual grass yield is 80t dry matter/ha, achieved by year-round growth in the tropics.* The yield involved the assimilation of over 1600 kg nitrogen/ha, almost all of which was supplied by fertilizer. As ruminants digest approx. 60% of the dry matter, this yield corresponds to about 50000 feed units or 50t grain equivalent/ha.

The maximum productivity of temperate perennial crops has been estimated at 30t dry matter/ha. † J. P. Cooper and E.L. Breese, in their paper "Plant Breeding:Forage Grasses and Legumes," a contribution to the symposium "Potential Crop Production," (London, 1971) write:

> The efficiency of existing forage varieties in converting the seasonal input of light energy, and hence their potential production, can be assessed from sward plots grown without the limitations of water or soil nutrients, and under a system of harvesting which maintains adequate light interception, but minimises senescence and decay at the base of the sward.
>
> In the maritime regions of Western Europe, such a system usually involves five to eight harvests during the growing season, a fertilizer input of up to 600 kgN/ha and corresponding amounts of phosphate and potassium, and the maintenance of an adequate water supply during the summer. With this type of management, highly tillering varieties of perennial ryegrass have revealed a potential of over 25 t/ha at Aberystwyth and up to 20 t/ha at Cambridge and in the Netherlands.

Professor W. Holmes has suggested that with a nitrogen input of 450 kg/ha/year, 13200 kg/ha dry matter should be available to the grazing animal during a 180-day grazing season. ‡

Mean intensities of nitrogen application to grassland, even in the most highly industrialized countries, are far below the limiting levels mentioned above, and are probably well below 100 kg/ha. National average green fodder yields are therefore correspondingly lower, as is shown in Table 20, which gives yields and areas for the EEC countries in 1973-4.

In the USA, the yield of hay and other harvested roughage on 65 million hectares of arable land in 1974 was 1500 feed units/ha; the green fodder yield on 244 million hectares of permanent grassland in the same year was 1150 feed units/ha. (Estimates based on USDA and OECD statistics).

*R. S. Loomis, "Agricultural Systems", *Sci. Am.* **235**, 1976, pp. 98-105.

†D. F. Westlake, "Comparisons of Plant Productivity", *Biol. Rev.* **38**, 1963, pp. 385-425.

‡The use of nitrogen in the management of pasture for cattle", *Herb. Abstr.* **38**, 1968, pp. 265-278.

Table 20.

Country	Green fodder yield (feed units/ha)	Rotation grass area (incl. clover and lucerne) (10^3 ha)	Permanent pasture area (10^3 ha)	Total green fodder area 10^3 ha)
German Fed. Rep.	4100	870	5264	6134
France	2700	4755	13555	18310
Italy	2200	2690	5214	7904
Netherlands	4950	121	1255	1376
Belgium-Luxembourg	4500	117	799	916
UK	2050	2353	11484	13837
Ireland	2700	815	3572	4387
Denmark	4100	470	277	747
EEC-9	2700	12190	41420	53610

Sources: EEC Feed Balance Sheet: Resources, 1976; EEC Yearbook of Agricultural Statistics, 1976.

R. C. Burns and R. W. F. Hardy estimate that 45 million tonnes of nitrogen is fixed annually on the world's meadows and grasslands, ("Nitrogen Fixation in Bacteria and Higher Plants," Springer, New York, 1975). As grass contains on average 30 feed units per kg nitrogen, world grass production is 1300 billion feed units, or 1.3 billion tonnes grain equivalent. The mean yield of the world's 3 billion hectares of grassland is therefore about 450 feed units/ha, or a quarter of the mean grain yield.

DESALINATION OF SEAWATER

Desalination of seawater has been suggested as a means of augmenting water supply for agriculture, but the energy requirement rules out its use on a significant scale. The energy requirement on a test run at the Chula Vista plant, near San Diego, California, was 31 kWh/m^3, equivalent to the energy needed to raise water through a height of 11.4 km. Supplying a depth of 750 mm of desalinated seawater would cost 28t coal-equivalent/ha; even if the resulting crop yield were 8t/ha, the energy input for desalination alone would be 7 times the energy content of the harvested grain. By way of comparison, the total on-farm energy input in U.S. maize cultivation in 1975 was approx. 45% of the energy content of the harvested maize.*

It is possible that improvements in desalination technology will result in a reduction of the energy requirement. The limit attainable under practical conditions has been estimated at 3.5 kWh/m^3, excluding the energy needed to raise the seawater to the level of the desalination plant.†

THE OZONE PROBLEM

In addition to the decline in the incremental yield–nitrogen ratio at high intensities of fertilizer application, there is a further reason why intensity of nitrogen application on cropland cannot increase indefinitely. A proportion of the nitrogen in the soil is converted to nitrous oxide (N_2O) and dissipated into the atmosphere, where part of it is photochemically converted to nitric oxides (NO and NO_2) which react with ozone (O_3) to convert it to oxygen (O_2). A reduction of the ozone layer would result in an increase in the intensity of deleterious ultraviolet radiation in the wavelength range 295-320 nm incident on the Earth's surface.

Global nitrogen fixation in 1974 has been estimated at 237 million tons, of which 39 million tons was contributed by chemical fertilizer. Estimates for global fixation in 1850 and 1950 are 150 million tons and 174 million tons respectively. Increasing the mean intensity of nitrogen application to 200 kg/ha on the world's billion hectares of cropland would mean an additional annual fixation of 200 million tons relative to the 1950 level. The effect of such an increase on the ozone layer is still highly uncertain – estimates of the reduction vary from 2% to 23%.‡ The National Academy of Sciences (USA) and others have estimated that a 10% reduction in the ozone layer would result in a 20% increase in the incidence of skin cancer.

*D. Pimentel and M. Pimentel, "Counting the kilocalories", *Ceres*, **10**, (5), pp. 17-21. FAO, Rome, Sept-Oct. 1977.

†A. A. Delyannis, "Nuclear Energy Centres and Agro–Industrial Complexes", IAEA, Vienna, 1972.

‡P. F. Pratt *et al.*, "Effect of increased nitrogen fixation on stratospheric ozone", Climatic Change, Vol.1, 1977, pp. 109-135

NEW PLANT DISEASES AND PESTS

The development of new high-yielding crop varieties is not without its dangers, even if the new varieties are resistant to known diseases and pests. The problem has been concisely stated by W. M. Myers, in his contribution to the symposium "Research for the World Food Crisis" (Washington, 1970):

> Continuous intensive agricultural research programs are required to maintain the advances made through previous research programs. We know that new diseases and insect pests may develop, capable of striking down the resistant varieties of crops now being used. This has been the experience in the temperate zone where, for example, new races of black stem rust of wheat have repeatedly arisen to which the once resistant varieties are susceptible. With this knowledge, one must view with concern the fact that millions of acres of wheat and rice are being planted in Pakistan and India and other countries to one or a small number of varieties of each crop. A disease organism capable of attacking one of these varieties could, in a short time, devastate vast areas and cause even more serious shortages of food than would have existed had these new superior varieties not replaced the old indigenous varieties. The only answer we know today to this threat is a continuous vigorous research program to develop further new varieties capable of replacing those in use when they become endangered by new hazards.

New plant diseases and pests are by no means the only danger, as Professor P. J. Newbould points out in his contribution to the symposium "Potential Crop Production" (London, 1971):

> Even if we learn within the next 100 years (and we are unlikely to do it sooner) to stabilize human population, agricultural production still needs to increase at least threefold. The way to do this may be by increased mechanization, chemical fertilizers, plant breeding, monoculture and pest control, in fact an acceleration of present trends, based as they are on comprehensive control of every aspect of an unstable and precarious system. Or these methods may turn sour on us. Chemical pesticides are creating a situation in which their use is bound to increase, and the chance of poisoning ourselves as well as the pests will also increase. Also the cost of pest control is increasing. It is not possible at present to predict whether or not the chemists will keep ahead of the pests. Also the widespread use of relatively heavy applications of inorganic fertilizers seems to be accelerating the eutrophication of fresh waters all over the world. We may need completely different techniques of cropping ecosystems.

THE CARBON DIOXIDE PROBLEM

The concentration of carbon dioxide in the atmosphere is increasing as a result of fossil fuel combustion, rising from 290 ppm in 1900 to 314 ppm in 1960, reaching 327 ppm in 1975, at which date the annual increase was 1 ppm. If fossil fuel combustion increases at 4% annually the concentration will reach 380 ppm by the year 2000 (Appendix E). A higher CO_2 concentration should make possible an increased efficiency of photosynthesis, provided that there are proportionate increases in the availability of nutrients (nitrogen, phosphorus, potassium etc). It is possible, however, that the concomitant increase in pollutants from fossil fuels will have an inhibitory effect.

Combustion of the world's measured reserves of fossil fuels (700 billion tons coal equivalent) would result in a CO_2 concentration of approx. 450 ppm. Combustion of the estimated total recoverable reserves (6000 billion tons coal equivalent) would result in a CO_2 concentration of 1300 ppm even in the unlikely event of vegetation and hydrosphere continuing to absorb 50% of the CO_2 emission. The danger involved is that rising CO_2 concentration will cause a rise in mean global temperature, as CO_2 molecules permit sunlight to pass down through the atmosphere, but reflect infrared heat waves emanating from the Earth's surface – the "greenhouse effect." It is true that the mean temperature of the northern hemisphere fell by 0.5°C in the period 1940-65. It has been suggested that this fall was caused by an increase in the amount of particulate matter (aerosols) in the atmosphere, the assumption being that aerosols increase the albedo (reflectivity) of the Earth, thereby reducing the amount of solar radiation incident on the surface. Since 1965, however, the mean temperature of the northern hemisphere has been rising, and by 1975 was 0.3°C above the 1965 figure.

Estimates of the rise in mean global temperature which would be caused by a doubling of the CO_2 concentration (to 600 ppm) vary from 0.7°C to 9.6°C, two of the most recent estimates being 1.5°C and 2.9°C. There is general agreement that any rise in mean global temperature would be amplified severalfold in the polar regions.*

Steadily rising temperatures would lead first to melting of the pack ice in the Arctic Ocean, and ultimately to melting of the inland ice of Greenland and Antarctica. Melting of the Greenland and Antarctic ice-

*S. H. Schneider, " On the Carbon dioxide–Climate confusion," *Journal of Atmospheric Sciences* 32, Nov. 1975.

sheets would result in world sea level rising by 7 m and 70 m respectively, submerging large areas at present densely populated and containing much of the best arable land. It is still too soon to say how seriously these dangers should be taken.

World sea level has risen by 200 mm since the beginning of the century, and is at present rising at a rate of 1.2 mm per year, corresponding to an annual increase of 430 km^3 in the volume of the oceans. This rise is not due to melting of the polar ice, but to a diminution in the volume of groundwater.

CONCLUSION

The considerations advanced in this chapter show that full utilization of current agro-technology would probably enable the mean world cereal yield to be raised to approximately 4.7 t/ha, or two and a half times the present yield. The projected yield limit should not be regarded as an absolute ceiling, but as a barrier which will be impossible to pass unless a substantial improvement in the efficiency of water and nitrogen utilization of the most important cereals can be achieved. The probability of such a breakthrough being made and utilized in the critical decades ahead is beyond the limits of conjecture. As regards non-cereal crops the position is less clear, but it seems likely that the mean global yield will remain close to the mean cereal yield.

It has also been shown that there are inherent dangers in moving towards complete dependence on agro-technology. It is quite possible that the population load on the world's cropland will ultimately have to be reduced if ecological equilibrium is to be maintained. For the next two centuries at least, however, there is no alternative to steadily increasing dependence on agro-technology, be the consequences what they may.

V

ARABLE AREA

The world's arable area is at present (1975) 1500 million hectares or 10% of the land surface. There has been much discussion in recent years concerning the possibility of augmenting this area. The view has frequently been expressed that the arable area could be increased to at least 2500 million hectares, chiefly by utilizing grassland and rain forests in Latin America and Africa. Though the cultivated area is increasing more rapidly in these regions than in the rest of the world, there is reason to believe that expanding the cultivated area in the tropics will not be of more than marginal significance for world food production. The following statements are cited in support of this view.

> The lands now in use for the production of crops are of certain specific types which are both fertile and relatively stable under present agricultural practices. The brown chernozem soils of the temperate regions and the red laterites of the less rainy tropics are the most fertile and the most easily cultivated areas. These soils are completely occupied by human beings and their agriculture at the present time. In order to gain more acres for our agriculture it will be necessary to expand the cultivated area into lands which are more difficult to use – the podzols of the wet cold regions and the laterities of the more rainy tropics. There are technical difficulties in the utilization of these lands, especially those of the wet warm regions. These soils are difficult and expensive to fertilize and to maintain in a state of good tilth because of the high leaching and erosion rates.
>
> H. Brown, J. Bonner, J. Weir: "The Next Hundred Years," (Viking, New York, 1957)

It is worth noting in connection with the statement quoted above that the total cultivated and cultivable area of the six most fertile soils (chernozems, chernozem-like prairie soils, grey forest soils, brown forest soils, red earths of the subtropics and alluvial soils) has been estimated by Soviet Academician I. P. Gerasimov, in a paper presented to the Royal Geographical Society in 1958, at 1244 million hectares. This area includes 98 million hectares of mountain soils.

A few years ago people looked to the day when the vast tropical forested areas would be used for food production. It is now thought that these tropical areas are apparently best suited for growing what they have always grown – trees. When cleared, the land rapidly loses its fertility and even with the addition of fertilizer little of this land is suitable for sustained production of food-crops. The advanced prehistoric cultures, with few exceptions, never developed in this kind of area and then, as now, the inhabitants of such areas depended on a slash-and-burn, shifting type of cultivation. Small areas are cleared and burned, crops are grown for a few years, and then the land is abandoned and a new area sought. Obviously, this type of cultivation will not support a large population. Although certain tree crops might be grown in these tropical forest areas, we can hardly count on them to contribute much to the world's food supply.

C. B. Heiser Jr. : "Seed to Civilization,"
(Freeman, San Francisco, 1973)

Once the lush forests are cleared, the soils of these regions are exceedingly vulnerable to the incessant rain and soon lose their fertility. In some cases the soil even hardens irreversibly to the texture of concrete, so that it is more suitable as a building material than as an agricultural resource (Cambodia's Angkor Wat is built of this material, called laterite). Few crops have been grown satisfactorily on a commercial basis in the humid tropics, and these tend to be low-protein or non-food cash crops such as palm-oil, sugar-cane, bananas, coffee, cocoa, and latex. Even the native crops tend to be low-protein varieties such as cassava, yams, and plantains. Here the invention of a completely new agriculture will be necessary, either to develop tree-like plants that yield ordinary nourishing foods and can withstand the hot and humid climate or to generate a series of plants that grow under the cover of tropical trees, which protect them and their soil against the pounding of the heavy tropical rains that otherwise cause destruction by horizontal erosion or vertical washouts. For decades to come only minimal contributions to the solution of the world's food problem can be expected from the humid tropical zones.

M. Mesarovic, E. Pestel: "Mankind at the Turning Point,"
(Hutchinson, London, 1975)

Competent international organisations have stated that the world's cultivable area is 24 million km^2. I cannot accept this figure, as in the tropics, where the bulk of the theoretical reserves are situated (Congo, Amazon), it is necessary to eliminate areas having a gradient exceeding one in a hundred; cultivation of areas with higher gradients than this would result in destruction of the soil by erosion within a few years. If forested areas of low fertility are also eliminated, we are left with barely 50% of the so-called potential arable area. The total arable area is certainly not greater than 20 million km^2, of which 14 million are cultivated. But these 14 million are the best lands of the world, since for thousands of years men have chosen them. (Transl. BG).

M. Guernier: "L'équilibre fragile," (Dossiers Jeune Afrique 2, Paris, 1975)

Not only is it very doubtful whether land to be brought into the cultivation cycle will be as productive as the land already under cultivation, but there is a serious risk, according to some agronomists, that soil erosion will ultimately reduce both the total arable area and its yield potential. Professor D. Pimentel considers it probable that the world arable area will be reduced by 30% over the next hundred years, chiefly as a result of erosion. He estimates the mean annual loss of topsoil in the USA at 30 t/ha. The mean annual rate of topsoil formation is 4 t/ha, so that the net loss is 26 t/ha. The thinner the layer of topsoil, the less is its capacity to retain water and nutrients, and the lower the crop yield for a given intensity of fertilizer application. It is true that eroded land is not irretrievably lost, but soil reclamation is a slow process; the highest attainable rate of topsoil formation has been estimated at 12 t/ha.

We are thus faced with the possibility of a loss of 500 million hectares of arable land by soil erosion and the encroachments of housing, industry and communications, and also the possibility of a gain of 500 million hectares by putting pastures and woodland under the plough. In making a projection for the next seventy years, it therefore seems reasonable to steer a middle course, and assume that land to be brought into the cultivation cycle will balance the loss due to erosion etc., so that the total amount of arable land remains at its present figure of 1500 million hectares.

VI
PROJECTIONS

In view of the many imponderables involved, it is impossible to make long-term projections of agricultural production, or even of cereal production, without a very wide margin of uncertainty. Many would say that the margin of uncertainty is so wide that no useful purpose is served by making such projections. I do not share this view. Long-term projections of food production, like those of population, are useful insofar as they are based on well-defined assumptions. The projections can then be revised from time to time in the light of new developments. This is precisely the procedure adopted by demographers for population projections.

A projection for the grain production of each of five major regions in the year 2050 is given below. These projections are combined with the "maximum" and "minimum" population projections for the same year, resulting in projections for per capita grain production. The grain projections are based on the following assumptions:

(i) That the intensity of application of chemical nitrogen on the cropped area will attain 200 kg/ha in all regions. (The implications of this assumption will be discussed in Chapter VIII).

(ii) That the incremental yield-nitrogen ratio in each region will remain at the mean 1963-73 value.

(iii) That the harvested grain area of each region will remain at the 1976 figure.

These assumptions, taken as a whole, are equivalent to the assumption of full utilization of the possibilities of current technology on the area at present harvested.

The population, grain production and harvested grain area of each region in 1976 are given in Table 21.

The projected grain yields, and the data on which they are based, are given in Table 22. (The values of Y_0 and the incremental yield–nitrogen ratio R are derived in Appendix D. The slight discrepancies between the regional classification used in Appendix D and that of Table 22 have been neglected.)

Table 21.

Region	P Population (10^6)	G Grain prod. (10^6 t)	A Harvest area (10^3 km^2)	G/P (kg/capita)	G/A (t/km^2)	P/A
DC	1100	770	3025	700	255	364
East Asia	906	259	1300	286	200	697
Latin America	333	87	541	262	161	616
Africa	413	70	712	169	98	580
South Asia	1292	291	2016	225	144	641
World	4043	1477	7594	365	194	532

Source: FAO Production Yearbook, 1976.

Table 22.

Region	Basic yield (Y_0) (kg/ha)	Incremental yield- nitrogen ratio (R)	Projected yield (Y_0+ 200R) (kg/ha)
DC	1210	16.3	4470
East Asia	1420	20.1	5440
Latin America	1170	17.0	4570
Africa	720	15.7	3860
South Asia	1040	19.7	4980

Tables 23 and 24 show population and per capita grain production for each region in 2050 on the "maximum" and "minimum" demographic projections respectively. The parameters given in the last three columns of Tables 21, 23 and 24 are shown graphically in Fig. 15.

Table 23.

Region	P Population (10^6)	G Grain prod. (10^6 t)	A Harvest area (10^3 km^2)	G/P (kg/capita)	G/A (t/km^2)	P/A
DC	1650	1352	3025	820	447	545
East Asia	2347	707	1300	301	544	1810
Latin America	1396	247	541	177	457	2580
Africa	1664	275	712	165	386	2340
South Asia	5916	1004	2016	170	498	2940
World	12973	3585	7594	276	472	1710

Table 24.

Region	P Population (10^6)	G Grain prod. (10^6 t)	A Harvest area (10^3 km^2)	G/P (kg/capita)	G/A (t/km^2)	P/A
DC	1420	1352	3025	952	447	470
East Asia	1863	707	1300	380	544	1430
Latin America	1077	247	541	230	457	1990
Africa	1244	275	712	221	386	1750
South Asia	4350	1004	2016	231	498	2160
World	9954	3585	7594	360	472	1310

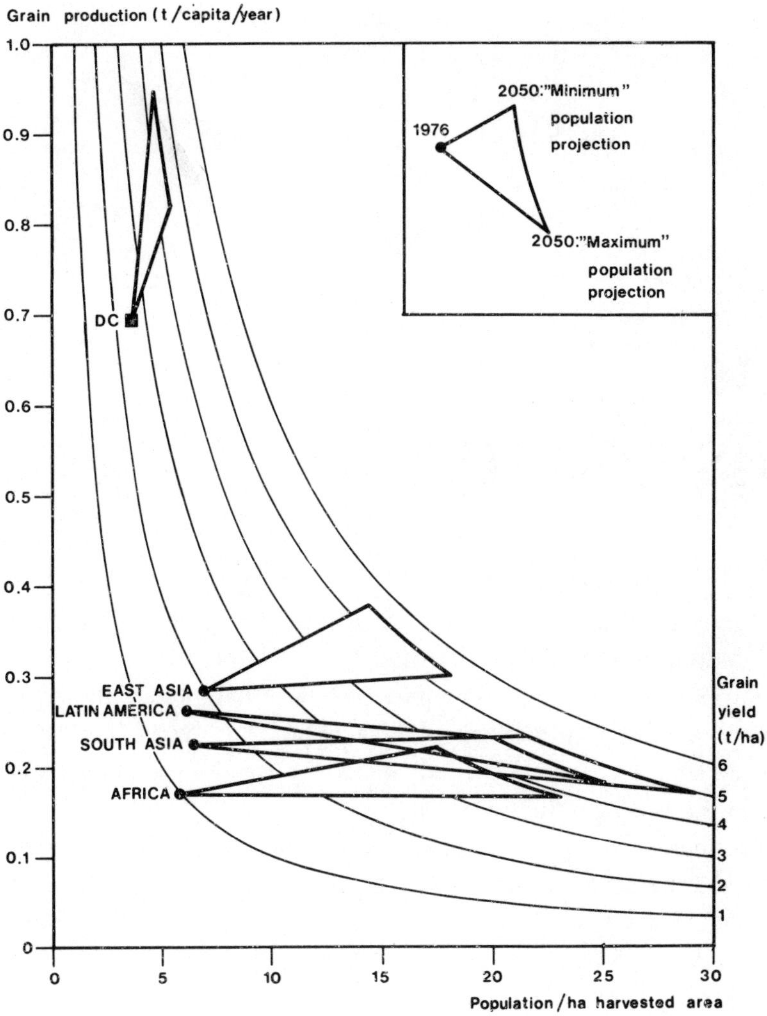

Fig. 15. Grain parameters, 1976–2050.

VII
DISPARITIES

In Table 21 three parameters – per capita grain production, grain yield, and population per km^2 harvested area – are given for each major region. Figure 16 shows the values of these parameters in 1973 for all countries which in that year had a population exceeding 18 million, a harvested grain area exceeding 35 thousand km^2, or a grain production exceeding 6 million tons.

An analysis of the agro-technological and crop production disparities between the DCs and the LDCs can be carried out by computing three parameter ratios, or disparity factors, for each of the principal categories of field crops. The three factors are

α = DC yield/LDC yield

β = DC harvested area per capita/LDC harvested area per capita = (LDC population per km^2 harvested area)/DC popu. per km^2 harvested area).

$\gamma = \alpha\beta$ = DC per capita production/LDC per capita production.

The values of α, β and γ for the principal field crops are tabulated below. The data are for 1973, and are taken from FAO Production Yearbook 1974. In accordance with the FAO classification, South Africa and Israel are included in the DC category.

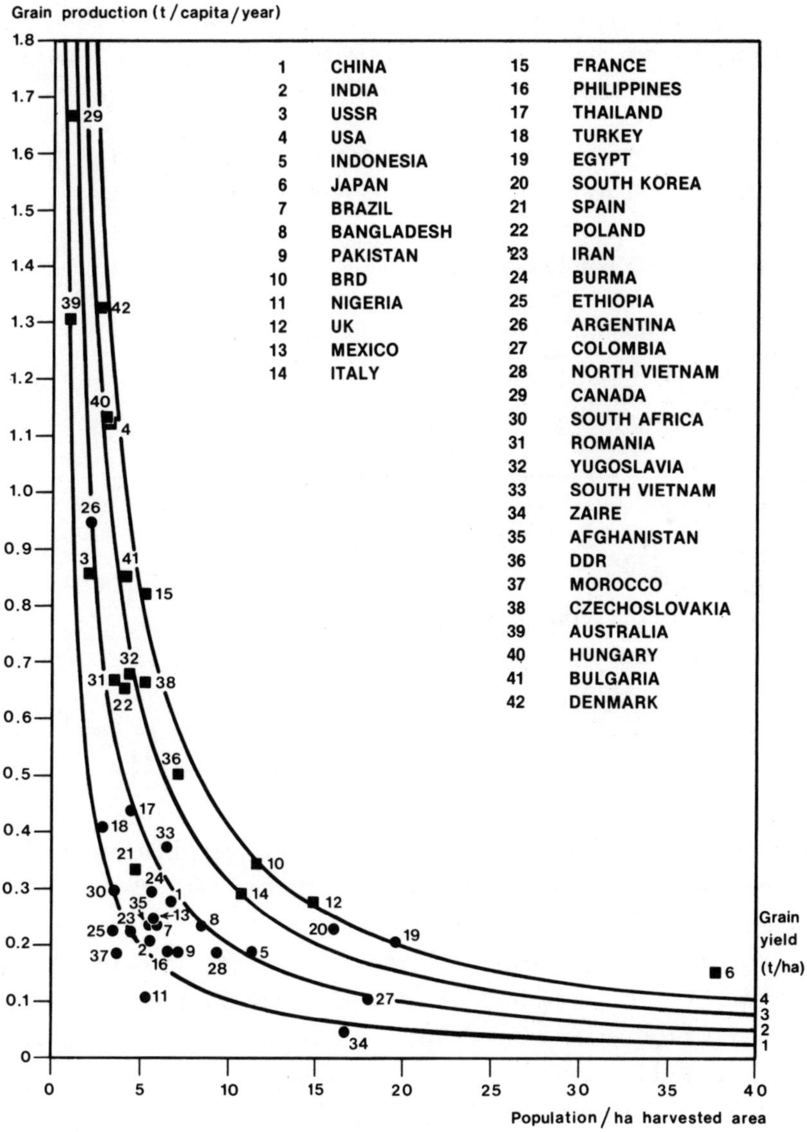

Fig. 16. Grain parameters, 1973.

(Source: FAO Production Yearbook 1974)

The production of any crop can be converted to grain-equivalent, i.e. grain of equal calorie content. Crop parameters and parameter ratios can then be computed for any combination of crops. The following conversion factors have been used for computing overall values of α, β and γ for the crops included in Table 25:

1 tonne roots and tubers = 0.2 tonnes grain
1 tonne soybeans = 1.2 tonnes grain
1 tonne pulses = 1.0 tonnes grain
1 tonne raw sugar = 1.1 tonnes grain

The overall values are found to be:

$$\alpha = 1.75; \quad \beta = 1.62; \quad \gamma = 2.82$$

Table 25.

	α	β	γ
Cereals	1.73	1.73	3.00
Roots and tubers	1.88	1.09	2.05
Soybeans	1.94	2.93	5.69
Pulses	1.86	0.53	0.98
Raw sugar	0.95	1.74	1.66

It can be seen that the overall disparity factors differ little from those based on cereals alone. It is clear that the present (1973) disparity between DCs and LDCs in per-capita crop production is due in almost equal measure to the disparity in techno-economic development and the pressure of population on cropland. Figure 17 shows per capita food energy production from the principal field crops and energy yields for the same 42 countries which are included in Fig. 16. Figure 17 presents much the same picture as Fig. 16, except for a few African countries in which production of root crops is high relative to that of cereals.

Projections for the disparity factors for cereals in 2050 can be derived from the data of Tables 23-24, and are given in Table 26.

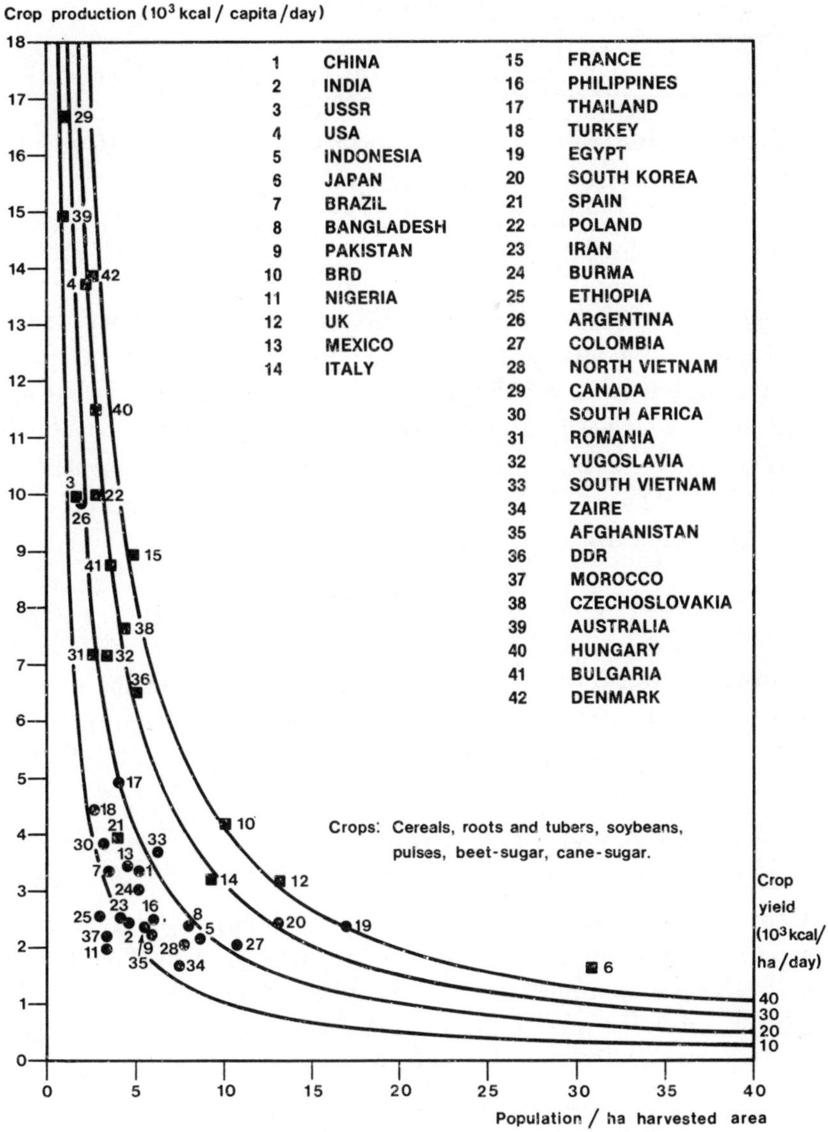

Fig. 17. Crop parameters, 1973.

(Source: FAO Production Yearbook 1974)

Table 26.

	α	β	γ
Maximum demographic projection	0.916	4.55	4.16
Minimum demographic projection	0.916	3.97	3.63

The projections developed in Chapter VI show that it is impossible for the LDCs to achieve anything approaching parity with the DCs in per capita grain production by the mid-21st century. The question then arises as to whether parity in per capita grain *consumption* is possible in the period under consideration. The net grain transfer from DCs to LDCs which would have been needed to equalize per capita consumption in 1976 can be seen from the data of Table 21 to be

$$1100\ (0.700 - 0.365)\ 10^6 = 368 \text{ million tons.}$$

This transfer is twelve times the net LDC grain import and about three times the net global transfer. The corresponding "equalization transfer" in 2050, based on the "minimum" population projection (Table 24), would be

$$1420\ (0.952 - 0.360)\ 10^6 = 840 \text{ million tons.}$$

Per capita grain production and net import or export of a number of countries are shown in Figs 18 and 19. It can be seen that per capita import in DCs varies inversely with per capita production, while that in LDCs is marginal. The reason for this is obvious enough. If, for example, India were to import 150 kg per capita, thereby raising consumption to the world average of 350 kg, the cost at current prices would be 30 dollars per capita. As India's per capita GDP in 1974 was 144 dollars, and per capita exports in the same year 7 dollars, such a volume of import is clearly impossible. Even if India were in a position to spend 30 dollars per capita on grain import, the sheer bulk of the increase in demand would result in a colossal price rise on the world market, so that the actual import would be much less than 150 kg.

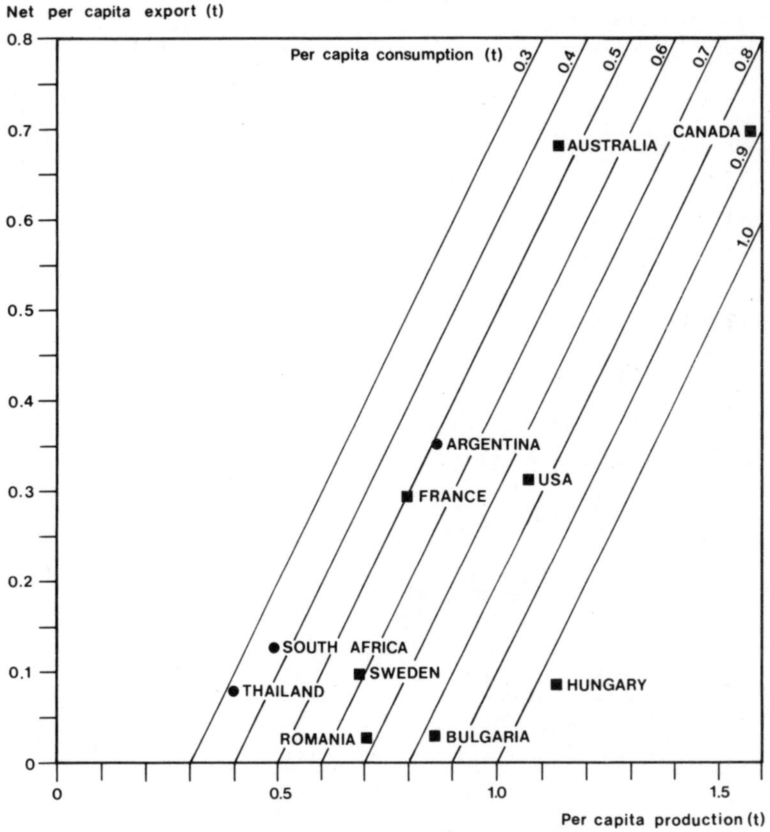

Fig. 18. Net grain exports. Per capita production and export, 1972–4 average. (Sources: FAO Production Yearbook, 1974 and FAO Trade Yearbook, 1974.)

The fact which must be faced is that grain transfer can be effected only by means of international trade, and that as long as the purchasing power of the average DC consumer is many times that of the LDC consumer, international grain trade will be dominated by the DCs. This leads to the question of whether it will be possible for the LDCs to approach parity with the DCs in per capita GDP. What will happen in the very long run is anyone's guess, but it is certain that parity by the middle of the 21st century is an economic impossibility, as this would necessitate extremely high rates of GDP growth in the LDCs and very low rates in the DCs throughout the intervening seventy-odd years. A certain narrowing of the development "gap" may occur, but barring

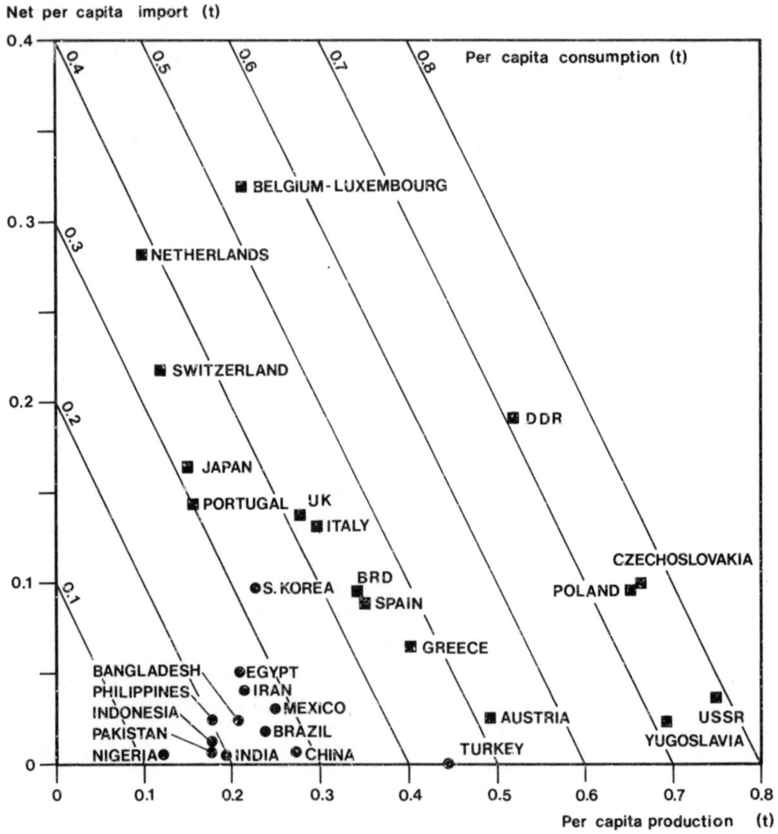

Fig. 19. Net grain imports. Per capita production and import, 1972–4 average. (Sources: FAO Production Yearbook, 1974 and FAO Trade Yearbook, 1974.)

major economic dislocations in the DCs, the gap in 2050 will still be large enough to prevent the net grain transfer approaching the amount needed to achieve a rough equality in per capita consumption. That this will be so can be shown by a consideration of the global energy situation.

The correlation between per capita GDP and per capita energy consumption is very high. The relation in 1972, as computed by J. Darmstadter, J. Dunkerley and J. Alterman ("How Industrial Societies Use Energy", Johns Hopkins Press, Baltimore, 1977), was

$$\log (E/P) = 0.082121 + 1.0048 \log (GDP/P)$$

where E/P = energy consumption per capita in tonnes oil equivalent (1 t oil equivalent = 1.42 t coal equivalent). and GDP/P = Gross Domestic Product per capita in thousands of US dollars, converted at market exchange rates.

(In computing energy consumptions the authors converted non-fossil electrical energy to oil equivalent on the basis of 35% conversion efficiency.)

Parity in per capita GDP between any two major regions clearly involves an approximate parity in per capita energy consumption, and this relation will almost certainly hold for a long time ahead, regardless of future changes in modes of energy utilization. The fact that some countries (such as the USSR, S. Korea and Czechoslovakia) use appreciably more energy to generate a dollar of GDP than the world average, and others (such as Brazil, France and Portugal) appreciably less, is of little significance in the context of the global development dichotomy.

The difficulties confronting the LDCs in reducing the disparity in per capita energy consumption between themselves and the DCs are aggravated by the differential in population growth rates and the high capital costs of developing non-fossil energy resources. The latter will be discussed in the following chapter.

A further difficulty is discussed by W. Häfele and W. Sassin, in "Energy Options and Strategies for Western Europe" (Science 200, p. 164-167, 14 April 1978). They estimate that a reasonable upper potential for global energy production in the year 2030 is 35 TW (38 billion tons coal equivalent). As world energy consumption in 1975 was 7.5 TW (8 billion tons coal equivalent), this potential implies that the mean global rate of growth of energy use over the fifty-five year period in question cannot exceed 2.8% per year. Häfele and Sassin consider that this will permit a global economic growth rate of not more than 3.7% per year, and that a realistic upper limit for global economic growth is 3.2% per year. They conclude:

> The dilemma is that the developing countries will demand growth rates higher than the global average. Corresponding growth rates lower than 3.2% for the developed countries would further aggravate their balance of payments problems, which result from increasingly expensive energy imports. Instead, they will try to stimulate economic growth in order to partly compensate for the rising share of energy costs in their balance of payments. In other words, it is possible that the north-south confrontation will become worse.

Häfele and Sassin's global energy limit implies an inverse relation between the maximum attainable long-term economic growth rates of the DCs and the LDCs. In terms of energy, a long-term growth rate of 3.2% in the DCs would involve zero growth in LDC energy consumption, while a long-term growth rate of 5.7% in the LDCs would be accompanied by zero energy growth in the DCs.

It is at all events clear that if the DCs and the LDCs maintain equal rates of energy growth, at no matter what level, the DC/LDC ratio in energy consumption per capita will increase from 11 to 1 in 1975 to at least 22 to 1 by the year 2030. A substantial differential in energy growth rates in the LDCs' favour will thus be necessary to prevent the development gap widening. Under these circumstances the prospect for parity is remote.

VIII
ENERGY

The pattern of global energy production and consumption differs from that of grain insofar as there is a net energy transfer from LDCs to DCs. The transfer amounts to 15% of world energy production, as can be seen from the data of Table 27. This situation arose during the 1960's with the rapidly increasing exploitation of Middle Eastern oil resources, and may well continue until the end of the century. It cannot continue much longer than this, if only because of resource depletion. The result of this phase of global development is that the OPEC countries, from the end of 1973 to the end of 1977, accumulated a capital of 180 billion dollars and enormously increased their imports. Though the rate of capital accumulation is slowly declining, the total will probably approach a trillion dollars by the year 2000, and enable several OPEC countries to complete the transition to industrial societies during the 21st century.

Per capita energy consumption and electricity production in 1975 in the sixteen post populous countries is shown in Table 28. It should be noted that though 1 t coal-equivalent corresponds to 8138 kWh of thermal energy*, the mean efficiency of conversion of heat to electricity in thermal power plants is 37%, so that only 3000 kWh of electrical

*7000 cal/g coal equivalent.

Table 27.

Region	Energy production (1975) (10^6 t coal-equiv.)	Energy consumption (1975) (10^6 t coal-equiv.)
DC	5241	6509
Latin America	420	311
Africa	456	158
South Asia	1776	363
East Asia	662	661
World	8555	8002

Source: UN Statistical Yearbook 1976.
Note: The UN statistics include energy from coal, lignite, petroleum, natural gas and hydro-electric, nuclear and geothermal installations; non-commercial energy sources such as wood, bagasse, vegetable refuse and cowdung are excluded. Hydro-, nuclear and geothermal electricity has been converted to coal equivalent on the basis of 1000 kWh = 0.123 t coal equivalent.

energy is produced per ton coal-equivalent consumed at the power plant.*

Measured recoverable reserves of fossil fuels in the major regions are given in Table 29, and the corresponding production figures for 1975 in Table 30. Total recoverable reserves of solid fuels may be ten times, and of crude petroleum and natural gas two to three times, the measured reserves.

World production of crude petroleum and natural gas in 1975 was 5.7 billion tons coal-equivalent (Table 30). If production were to continue at this rate, the measured reserves would be exhausted by 2010. Production will very probably begin to decline before this date, even if

*The mean conversion efficiency in the OECD countries in 1975 was 36.6%. The highest national average efficiency was that in the Netherlands – 41.7%. (Energy Balances of OECD countries 1973/1975, Paris, 1977). The replacement of steam turbines and dynamos by magnetohydrodynamic generators, in which heat from ionized gases is directly converted to electricity, may ultimately enable an efficiency of 60% to be achieved.

substantial additional reserves are discovered. *It is clear from Tables 29 and 30 that the DCs and East Asia will be able to maintain a substantial per capita production of solid fuels during the transition to non-fossil fuel energy sources, and that none of the four LDC regions will ever be able to attain a per capita fossil fuel production equal to that of the DCs.

Table 28.

Country	Energy consumption (t coal-eqv/cap.)	Electricity production (kWh/cap.)	Country	Energy consumption (t coal-eqv/cap.)	Electricity production (kWh/cap.)
DC			LDC		
USSR	5.55	4083	China*	0.56†	147†
USA	11.00	9367	India	0.22	143
Japan	3.62	4288	Indonesia	0.18	25
BRD	5.34	4881	Brazil	0.67	729
UK	5.26	4864	Bangladesh	0.03	18
Italy	3.01	2608	Pakistan	0.18	125
France	3.94	3382	Nigeria	0.09	51
			Mexico	1.22	720
			Philippines	0.33	291

*Excluding Taiwan. †Chu-yuan Cheng, "China's Energy Resources", Current History, 74, March 1978, pp. 121-124.
Source: UN Statistical Yearbook., 1976.

Though fossil fuels will continue to be the principal energy source for many decades to come, it is obvious that other sources must be developed in order to replace them as reserves are depleted. The problem confronting the LDCs is that development of alternative energy sources is capital-intensive, and also that the utilizable potential of several of these sources is limited. The alternatives will be briefly reviewed.

*According to projections developed by the Workshop on Alternative Energy Strategies, world production of crude petroleum will peak by 2004 at the latest. The assumptions on which the projections are based are given by A.R. Flower in "World Oil Production", Scientific American 238, No. 3, Mar. 1978. p. 42-49.

Table 29.

Region	Solid fuels (10^9 tcoal-eqv)	Crude Petroleum (10^9 tcoal-eqv)	Natural gas (10^9 tcoal-eqv)	Total (10^9 tcoal-eqv)
DC	408	24	40	472
Latin America	2	12	3	17
Africa	13	19	7	39
South Asia	12	77	15	104
East Asia	81	3	1	85
World	516	135	66	717

Source: World Energy Conference, Survey of Energy Resources, 1974.

Table 30.

Region	Solid fuels (10^6 tcoal-eqv)	Crude petroleum (10^6 tcoal-eqv)	Natural gas (10^6 tcoal-eqv)	Total (10^6 tcoal-eqv)
DC	1914	1646	1502	5062
Latin America	13	338	51	402
Africa	75	358	18	451
South Asia	109	1575	82	1766
East Asia	529	118	5	652
World	2640	4035	1658	8333

Source: UN Statistical Yearbook, 1976.

(i) Hydro-Electricity

The utilizable hydro potentials of the five regions, and the per capita potentials in 2050 (based on the "minimum" population projection) are given in Table 31.

Table 31.

Region	Hydro capacity 1976 (GWe)	Potential hydro capacity (GWe)	Population 2050 (10^6)	Potential per capita hydro capacity (kWe)
DC	309	861	1420	0.606
Latin America	24	328	1077	0.305
Africa	10	437	1244	0.351
South Asia	18	296	4350	0.068
East Asia	11	339	1863	0.182
World	372	2261	9954	0.227

Source: World Energy Conference, Survey of Energy Resources, 1974.
World Energy Supplies 1972-1976, UN, New York, 1978.

In a report presented to the World Energy Conference at Istanbul in September 1977, it is estimated that the cost of increasing the world hydro-electric generating capacity from 16% of the potential to 80% (i.e. from 372 GWe to 1820 GWe) would be 1450 billion dollars, corresponding to 1000 dollars/kWe. The unit cost of hydro-electric installations is subject to considerable variation, depending on the topography of the site. The 12.6 GWe Itaipu installation on the Parana River (Brazil-Paraguay), due for completion in 1989, will cost only 650 dollars/kWe, as the location is exceptionally favourable.

It can be seen from Table 31 that even full utilization of the LDCs' hydro potential would not suffice to bring per capita electricity production up to the current level in the DCs, where the total (thermal, hydro and nuclear) generating capacity is over 1 kWe per capita. (The mean load factor being close to 0.5 in almost all countries, annual electricity production per kWe is approx. $0.5 \times 24 \times 365 \approx 4400$ kWh).

(ii) Nuclear Energy

The capital cost of nuclear power plants in 1977 was 800 dollars/kWe, considerably more than that of oil-fired power plants (400 dollars/kWe) or coal-fired plants (450 dollars/kWe). Construction of nuclear capacity sufficient to provide 1 kWe per capita in Latin America, Africa and

South Asia by the year 2050 would run to over 5000 billion dollars even on the assumption of the optimistic "minimum" population projection.

The growth of nuclear generating capacity is impossible to estimate with any confidence more than a few years ahead, as it will depend on many imponderables, including the future volume and pattern of uranium production and trade, and the extent to which breeder reactors replace or supplement uranium reactors. Present trends indicate a capacity of 1000 GWe by the year 2000.*

Reserves of uranium in the non-Communist countries, recoverable at a cost not exceeding 130 dollars/kg, are estimated at 4.3 million tonnes, corresponding to 127 billion tonnes equivalent. The bulk of the reserves are located in USA, Canada, Australia and South Africa ("Uranium-Resources, Production and Demand," OECD/IAEA, Dec.1977).

Breeder reactors may ultimately enable up to sixty times as much energy to be produced from a given amount of uranium as can be produced in "conventional" uranium reactors. The latter produce 240 GWh of thermal energy, or 29500 tonnes coal equivalent, per tonne uranium, which is converted to electricity at the same efficiency as in fossil-fuel power plants – 35 to 40%.

World nuclear generating capacity in May 1978 was approximately 110 GWe, of which the LDCs accounted for 2 GWe. The total capacity of nuclear generators in operation, under construction, or on order is approx. 440 GWe, of which the LDCs (mainly Iran, South Korea, Taiwan and Brazil) account for 30 GWe.†

Fusion energy seems to offer unlimited possibilities, the theoretical energy reserve contained in the deuterium of the hydrosphere being 5×10^{20} t coal equivalent (Appendix F). It is highly problematic, however, whether fusion energy will ever be a practical proposition. It will in any case be capital-intensive; the cost per installed kW will certainly not be less than for fission generators, and may well be much more.

*"Nuclear Fuel Cycle Requirements and the Long-Term Outlook for Supply", *OECD Observer*, 92, May 1978, pp. 18-22.

†I am indebted to Mr. F. List of the Risø Experimental Station, Denmark, for this information.

(iii) Solar energy

Electricity production by direct conversion of solar radiation will be even more capital-intensive than nuclear energy. A 10MWe "power tower" solar thermal generator near Barstow, California, construction of which is to begin in 1978, will cost 12000 dollars/kWe; a 1 MWe "power tower" to be constructed in Italy, and scheduled to be operational in 1981, will cost 9000 dollars/kWe. A. F. Hildebrandt and L. L. Vant-Hull estimate that 100 MWe solar tower plants could be mass-produced at 1700 dollars/kWe ("Power with Heliostats", Science 197, 1977, p.1139-46).

The geosynchronous satellite solar power station, a proposal first advanced by Dr. P. E. Glaser in 1968, is estimated to cost 1500 dollars/kWe, almost half of which is the cost of transportation to orbit. The estimate is based on the assumption that the power-weight ratio of photo-electric cells can be increased to many times the present ratio ("Solar Power from Satellites," U. S. Govt. Printing Office, Washington DC, 1976).

(iv) Wind energy

As the output of a wind-powered generator varies with the cube of the wind velocity, it is subject to continuous wide-range fluctuation. Wind power is therefore unsuitable as a principal source of electrical energy, and will remain so until energy storage technology is much more advanced than it is at present.

A wind-powered generator ordered by ERDA from Boeing, due for completion in 1979, will have a 91.5 m diameter rotor and is rated at 2500 kWe. The contract price is 10 million dollars, or 4000 dollars/kWe. The two wind-powered generators under construction at Velling Maersk, Denmark, will have 40 m diameter rotors and a rated capacity of 600 kWe·each. The estimated annual electricity production is 1.5×10^6 kWh per generator, corresponding to a load factor of 0.3. The cost is 1.3 million dollars per generator, or about 2200 dollars/kWe. The effective unit cost, based on the global average load factor of 0.5, is thus approx. 3700 dollars/kWe.

The unit cost of wind-powered generators could no doubt be substantially reduced if they were to be produced in large numbers, but it is very improbable that it will ever be competitive with hydro-electric or nuclear installations.

(v) Ocean thermal energy

The theoretical potential of this energy source has been estimated at 70000 GWe, but it is considered unlikely that more than 700 GWe is utilizable, power production by this method being feasible only in a small number of offshore areas.* A recent estimate of unit cost is 2500 dollars/kWe.

(vi) Geothermal energy

Utilization of all known geothermal sites that provide natural steam would provide approx. 60 GWe.* The world geothermal generating capacity in 1975 was 1307 MWe, of which USA accounted for 559 MWe, Italy 421 MWe, and New Zealand 192 MWe.

Harnessing the heat energy stored in the dry rocks of the Earth's crust is a different matter altogether, the technology of which is still in its infancy. Even if it ultimately became feasible, however, large-scale utilization would create formidable ecological problems.†

(vii) Tidal energy

The total tidal power on Earth is approximately 3300 GW, but economic utilization of tidal energy is possible only in bays and estuaries where the tidal amplitude is several metres. Even in these locations unit costs are higher than for hydro-electric installations based on river-flow.

Utilization of tidal energy at all practical sites in the world would provide about 65 GWe. Almost all these sites are located in N. America, W. Europe, USSR and Australia. The only tidal power plants in operation at present (1978) are at La Rance, near St. Malo, France (240 MWe), and Kislogubsk, near Murmansk, USSR (0.8 MWe). The projected tidal installations at Tugursk and Penzhinsk, on the Sea of Okhotsk, with potential capacities of 10 GWe and 35 GWe respectively, may

*"Survey of Energy Resources." New York, 1974.

†G. Lehner, Possibilities of Utilisation of New Non-Nuclear and Non-Fossil Energy Sources", *Kerntechnik,* **20**, No. 4 (1978), pp. 157-168.

never be fully developed, due to the remoteness of the locations from industrial areas.

(viii) Biomass energy

Wood is far and away the most important source of biomass energy, and is likely to remain so. Global fuelwood production cannot, however, increase much above the present level, as a major increase would result in rapid deforestation. As wood will be required in increasing amounts for other purposes (newsprint, construction material, plastics, furniture) a gradual decline in world fuelwood production is probable.

World production of fuelwood in 1975 is estimated at 1174 million cubic metres, of which the DCs accounted for approx. 150 million, Latin America 225 million, Africa 275 million, South Asia 370 million and East Asia 150 million.* (The average energy content of fuelwood is approx. 0.3 tonne coal equivalent/m^3). The principal fuelwood producing countries are Brazil (140 million m^3), China (137 m), India (113 m), Indonesia (111 m), USSR (83 m) and Nigeria (62 m). (Source: "World Energy Supplies 1971-1975", UN, New York, 1977).

Large-scale conversion of low-protein crops such as cassava and sugar cane to industrial alcohol has little prospect of making a significant contribution to energy production, as few countries will be able to spare the arable land. Even the 550 thousand tonnes of alcohol obtained from sugar cane in Brazil in 1977 provided only 0.6% of that country's commercial energy. Alcohol production in Brazil is scheduled to increase considerably during the next few years, but as the cost of production is higher than that of imported petroleum, the alcohol programme has to be subsidized. Conversion of Brazil's entire raw sugar production of 7.5 million tonnes would provide about 3.8 million tonnes of alcohol, corresponding to the same amount in coal equivalent. The total area planted to sugar cane in Brazil in 1976 was 2.24 million hectares.

A proposed source of biomass energy in developed countries is the conversion of whey to alcohol. The contribution of whey-alcohol to

*World roundwood production (including fuelwood) in 1975 was 2448 million m^3. J. Weck has estimated that full utilization of world forest resources would bring production up to 5228 million m^3 ("Die Wälder der Erde", Verständliche Wissenschaft 67, Berlin, 1957).

the energy supply will be extremely modest: If all the whey produced in the EEC in 1974 had been converted to alcohol, the resulting fuel energy would have been 400 thousand tonnes coal equivalent or about 0.03% of EEC energy consumption.

Conclusion

The upshot of this review of energy costs and potentials is that future economic development in most LDCs will be subject to severe constraints. The LDCs may nevertheless attain a degree of development which will enable them to fully utilize agro-technology by the middle of next century. What this will involve in terms of energy for chemical nitrogen fixation is considered below.

Energy and nitrogen fixation

Achieving crop yields in the region of 5 t/ha involves increasing the application of chemical nitrogen fertilizer to 200 kg N/ha per year, as shown in Chapter IV. As the energy requirement for nitrogen fixation is 2.7 tons coal equivalent per ton nitrogen,* the energy needed to produce the nitrogen will be approx. 0.5 t coal equivalent per hectare cropland. The world area cropped annually is close to a billion hectares, so that if all regions were to practise intensive cultivation, over 500 million tons coal equivalent would be needed for nitrogen fixation alone.

Table 32 shows nitrogen production, nitrogen consumption, and total energy consumption in the principal nitrogen producing countries. It can be seen that Pakistan and Egypt have by far the highest nitrogen/energy consumption ratios, Pakistan producing the bulk of its nitrogen and Egypt importing it. It can also be seen that Pakistan uses 7% of its total energy supply for nitrogen fixation, a higher proportion than in any other country. Whether this proportion will be attained or exceeded by other countries remains to be seen.

*Professor D. Pimentel's estimate ("Counting the Kilocalories", Ceres 10, No. 5, Sept.-Oct. 1977). G. Leach ("Energy and Food Production", Guildford, 1976) uses a figure of 80 MJ/kg N, corresponding to 2.74 t coal equivalent per t N. G. Leach and M. Slesser have estimated the theoretical minimum energy input for ammonia production at 36.9 MJ/kg; as the nitrogen content of ammonia (NH_3) is 82.4%, this corresponds to 1.53 t coal equiv./t N. The energy input for ammonia produced by ICI in 1973 was 50.5 MJ/kg.

Table 32.

	A Nitrogen production 1975-6 (10^3 t)	B Nitrogen consumption 1975-6 (10^3 t)	C Energy consumption 1975 (10^6 tcoal-eqv)	A/C (kg/t)	B/C (kg/t)
World	43876	43301	8002	5.5	5.4
USA	9262	9385	2350	3.9	4.0
USSR	8465	7357	1411	6.0	5.2
China	3300	4555	570	5.8	8.0
Japan	1557	638	402	3.9	1.6
Poland	1533	1224	170	9.0	7.2
India	1508	2031	132	11.4	15.4
France	1354	1708	209	6.5	8.2
Romania	1292	788	80.8	16.0	9.8
BRD	1259	1228	330	3.8	3.7
Netherlands	1153	453	79.0	14.6	5.7
UK	1055	1045	295	3.6	3.5
Italy	1000	724	168	6.0	4.3
Bulgaria	672	380	41.7	16.1	9.1
Mexico	581	833	73.4	7.9	11.3
S. Korea	541	468	36.0	15.0	13.0
Pakistan	316	442	12.8	24.7	34.5
Indonesia	208	342	24.2	6.1	14.1
Brazil	160	410	71.8	2.2	5.7
Egypt	150	415	15.1	9.9	27.5

Source: UN Statistical Yearbook, 1976.

It is possible that plant geneticists will one day develop new varieties of cereals which fix their own nitrogen by means of symbiotic micro-organisms, in the same way as soybeans and other dry legumes. There is, however, little prospect of such a breakthrough being made before next century, and a further two or three decades at least would be needed before the new varieties could replace the traditional ones on a global scale. It is by no means certain that symbiotic cereal plants would be as resistant to pests and diseases as traditional varieties. Most important of all, there is no reason to believe that biological nitrogen fixation in cereals would result in higher yields than those obtainable by means of chemical nitrogen fertilizer. Increased symbiosis offers at best a hope of a gradual reduction in the global demand for chemical nitrogen fertilizer.

IX

CONCLUSION

The demographic outlook for the DCs is that population in the year 2050 will be at most 50% greater than in 1976. As food production in the DCs can easily be increased by this percentage over the next seventy years, provided consumers are willing to pay more for their food, there is no reason to fear a decline in per capita food availability in these countries.

The population of the LDCs in the year 2050 will be nearly three times the 1976 figure even if the two-child family is achieved in East Asia by 2000 and in South Asia, Africa and Latin America by 2020. It is possible that a few LDCs will achieve the two-child family earlier than assumed in this timetable, but extremely improbable that the majority will be able to do so.

As almost all the best land is already under cultivation, additional land which may be brought into cultivation in the period up to 2050 will contribute little to the LDCs' food supply. Even this contribution may be more than counterbalanced by erosion losses on the land at present cultivated. The LDCs will therefore have to increase crop yields in the same ratio as population increase, on average from 1.5 to over 4 tonnes grain equivalent per hectare, in order to maintain current levels of per capita food production. This is technologically possible, but whether existing political and economic constraints will be relaxed sufficiently to permit the utilization of the necessary technology depends on factors which are unforeseeable.

LDCs which fail to maintain both a rapid rate of fertility decline and a rapid transition from traditional agriculture to modern agro-technology will experience a decline in per capita food production. In countries where the shortfall cannot be supplied by increased imports, the population will stabilize by the death-rate rising to the level of the birth-rate.

Food availability per capita in the LDCs could be increased substantially if these countries were in a position to increase grain imports to a level which resulted in a roughly uniform per capita grain consumption in all countries. Redistribution on this scale cannot be achieved without approximate parity in per capita GNP between DCs and LDCs. What parity implies in terms of energy has been discussed in Chapters VII and VIII. In economic terms, parity by the year 2050 would involve a mean annual GNP growth rate in the LDCs of 6% even if per capita GNP in the DCs stabilized immediately. A growth rate of this magnitude would involve gross fixed capital formation in the LDCs being rapidly increased from less than 20% of GNP to over 30%, and thereafter maintained at the latter figure. Theoretically, this could be achieved by the DCs supplying the investment deficit in the form of development assistance. There is evidence, however, that increasing the flow of foreign capital (public or private) into an economy leads to a lower rate of domestic saving.* If development assistance rose to 10-15% of the recipients' GNP, domestic capital investment would probably be greatly reduced, and the resulting rate of GNP growth be little higher than it would have been without any foreign aid at all.

It has been suggested† that the solution to global development problems lies in drastic cuts in military expenditures and the reallocation of the resources thereby saved in the DCs to development assistance. In addition to the probable negative effects of a massive increase in foreign aid on domestic investment, there are three insuperable objections to

*K. Griffin, "Underdevelopment in Spanish America," Allen and Unwin, London, 1969, p. 117-148.

†"Economic and social consequences of the arms race and of military expenditures," United Nations, New York, 1978.

this proposal: Firstly, the Soviet government cannot be trusted not to circumvent any disarmament agreement, acquire overwhelming military superiority, and use it to subjugate other countries. Secondly, it is inconceivable that the inhabitants of the industrialized countries, capitalist or Communist, would ever be prepared to sacrifice economic growth in order to donate several trillion dollars to the LDCs. Thirdly, the infrastructure in most LDCs is altogether inadequate to enable them to effectively utilize a rapid increase of development assistance to many times the present flow.

A world disarmament agreement is just as chimerical as a world agreement on population limitation, and ultimately for the same reason—it would be impossible to get immediate concerted military action by other countries if one country should violate the agreement to a slight extent. The result would be further violations, and it would not be long before the ball would be rolling again.

The upshot of these considerations is that the global dichotomy will be with us for a long time yet—certainly well beyond 2050. This means among other things, that the per capita food imports of the LDCs will be marginal. As the great disparity between DCs and LDCs in per capita crop production is certain to persist, the difference in the dietary patterns between the two groups of countries will also persist. A liberal supply of livestock products in national average diets will not be possible outside the temperate zones until long after the mid-21st century. It may never be possible at all unless population stabilization in the tropics is followed by a substantial population decline.

Finally, it should be emphasized that there is no ground for complacency regarding the future of the DCs. Both the developed market economies and the centrally planned economies of the Soviet bloc face many serious problems. The Malthusian threat, however, is not one of them.

APPENDICES:

Appendix A

CONVERSION OF FEEDSTUFFS TO LIVESTOCK PRODUCTS

Three analyses of the feedstuff input–livestock product output ratio will be made:

I. A complete analysis is given for the EEC (European Economic Community), the members of which are the only countries for which feed balance sheets are available.

II. Estimates are made for the cereal component of feed consumption and for the net production of animal protein in the USA, Canada, USSR, Japan, Australia and Argentina.

III. An estimate of the global food and feed balance in 1974, and a projected balance for 2050, will be given.

1. THE EEC

The statistics in this section are derived from the EEC Yearbook of Agricultural Statistics (1976), the EEC Feed Balance Sheet: Resources (1976), and the FAO Production Yearbook (1975).

Input

Feedstuff consumption in the EEC for the period 1 July 1973–30 June 1974 is summarized in the following table. The feed value of each category of feedstuff is given in feed units, one feed unit being the

Table 33.

Feedstuff	Feed value (10^6 feed units)	Crude protein (10^6 kg)
A. *Marketable feedstuffs*		
Cereals	73234	7003
Pulses	521	136
Potatoes	1977	163
Sugar	51	0
Vegetable oils and fats	605	0
Processed green fodder	1019	316
Manioc	1855	32
Other vegetable feedstuffs	264	37
By-products of the milling industry	6502	1155
By-products of the brewing, distilling and starch industries	1817	585
By-products of the sugar industry	4898	559
Soybean cakes	9361	4195
Other oilcakes	5146	1983
Other by-products of vegetable origin	535	63
Processed products of marine origin	714	487
Processed products of animal origin	763	632
Animal oils and fats	991	0
Milk and dairy by-products	6087	1150
B. *Feedstuffs not normally marketed*		
Root crops (green and as silage)	8295	927
Green maize (green and as silage)	10000	1192
Other green fodder crops (green and as silage)	4429	924
Products of meadows and grassland (green, as hay or as silage)	130655	26583
Straw and chaff	3533	406
Leaves and tops of root crops and vegetables	2897	656
Other crop by-products	421	98
Diverse unspecified feedstuffs	337	25
A + B	276907	49307

energy value of 1 kg barley. (The average feed value of cereals is very close to unity, i.e. 1 feed unit/kg).

The crude protein content of each category of feedstuff is also given. The crude protein content has been calculated by multiplying the measured nitrogen content by 6.25. It is therefore an estimate of the total amount of nitrogenous matter in the feedstuff, which is significantly greater than the amount of digestible protein.

Output

Production of animal protein in the EEC in 1974 is tabulated below, the protein content being estimated on the basis of an assumed protein content for each category of livestock product.

Table 34.

Product	Production (10^6 g)	Protein content (%)	Protein (10^6 kg)
Beef and veal	6483	17.5	1135
Pigmeat	8435	12.0	1012
Mutton, lamb, goatmeat	499	15.5	77
Poultrymeat	3152	20.0	630
Edible offals	1270	20.0	254
Whole milk	101896	3.5	3566
Eggs	3736	12.5	467
Gross production	–	–	7141
Milk and milk products recycled as feed			1150
Eggs for hatching			26
Net production	–	–	5965

Input/output ratios

Feedstuff/net animal protein = 276907/5965 = 46 feed units/kg.

Crude protein/net animal protein = 49307/5965 = 8.3

Feed grain/net animal protein = (73234+6502)/5965
= 13.4 feed units/kg.

The energy content of the feedstuff consumed

$$= 276907 \times 10^6 \times 3320 = 920 \times 10^{12} \text{ kcal.}$$

Energy contents of the livestock product outputs are calculated below:

Table 35.

Product	Production (10^6 kg)	Energy content (kcal/kg)	Total energy (10^{12} kcal)
Beef and veal	6483	1680	10.9
Pigmeat	8435	3960	33.4
Mutton, lamb, goatmeat	499	2410	1.2
Poultrymeat	3152	1290	4.2
Edible offals	1270	1430	1.8
Whole milk	101896	650	66.3
Eggs	3736	1440	5.4
Gross energy production			123.2
Milk recycled	6087×10^6 feed units		20.2
Eggs for hatching	5.6% of gross production		0.3
Net energy production			102.7

Feedstuff energy/gross livestock product energy = 920/123.2 = 7.5

Feedstuff energy/net livestock product energy = 920/102.7 = 9.0

Mean energy content of livestock products per unit protein

$$= (102.7 \times 10^{12})/(5965 \times 10^9)$$
$$= 17.2 \text{ kcal/g}$$

Population of the EEC countries in 1974 = 258.34 million

Feed grain per capita = 79736/258.34 = 308 feed units

Animal protein production per capita = 5965/258.34 = 23.1 kg/year
= 63 g/day

Total feedstuff consumption per capita

$$= 276907/258.34 = 1070 \text{ feed units/year}$$
$$= 9730 \text{ kcal/day}$$

II. ESTIMATES FOR OTHER COUNTRIES

As the EEC cannot be taken as representative of world agriculture, it is necessary to estimate input/output ratios for other countries. The following have been selected: USA, Canada, USSR, Japan, Australia and Argentina.

Input

We shall confine ourselves to estimating the cereal component of the total feedstuff consumption, no information on other feedstuffs being available.

The per capita feed grain consumption will be taken as

$$0.92A - B - C$$

where A = grain production per capita, B = net grain export per capita, C = net cereal consumption by humans, per capita.
The factor 0.92 is based on the assumption that grain for seed and industrial uses (brewing, distilling etc.) amounts to 8% of gross production.*

The following statistics are averages for the period 1973-5, taken from FAO Production Yearbook 1975 and FAO Trade Yearbook 1975. Statistics of cereals for human consumption are taken from FAO Production Yearbook 1971.

The figures given in the final column of Table 36 include the by-products of the milling industry. As the extraction rate for both flour and rice milling is approx. 75%, and virtually all the bran, husks etc are used for feed, the by-products of milling amount to one third of the net cereal consumption of humans.

*A reasonable approximation. Grain for seed and industrial uses in 1973 amounted to 6% of production in USA, 8% in Canada and 12% in Australia. ("Food Consumption Statistics 1955-1973", OECD, Paris, 1975).

Table 36.

Country	A (kg/cap.)	0.92A (kg/cap.)	B (kg/cap.)	C (kg/cap.)	Feed grain (kg/cap.)
Canada	1548	1424	621	66	737
USA	1086	999	357	64	578
Argentina	940	865	383	94	388
Japan	153	141	–170	128	183
Australia	1264	1163	650	79	434
USSR	708	652	–40	156	536

The margin of error in the feed grain estimates is chiefly due to neglecting variation in grain stocks.

Output

Per capita production of each category of animal protein is shown in Table 37. The figures for 1975, taken from FAO Production Yearbook 1975. The same conversion factors for protein content have been used as in the section on EEC output.

The figures for net milk protein are the gross production less estimated by-products used for feed. The bulk of the by-products of butter production (skimmed milk and buttermilk containing 3.5% protein) and all the by-products of cheese production (whey containing 1% protein) are used as feed. It has been assumed that 75% of the by-products of butter production and 100% of the by-products of cheese production are recycled, that 23 kg of whole milk are required to produce 1 kg of butter, and that 7 kg of whole milk are required to produce 1 kg of cheese.

Input/Output Ratios

Per capita consumption of feed grain, per capita production of meat, milk and egg protein, and the ratio of the former to the latter, are shown in Table 38.

The wide range of variation in feed grain-protein ratio is chiefly due to variation in the proportion of the total feed constituted by grass and

Table 37.

Country	Beef and and veal protein (kg/cap.)	Mutton and lamb protein (kg/cap.)	Pigmeat protein (kg/cap.)	Poultrymeat protein (kg/cap.)
Canada	8.35	0.05	2.77	4.54
USA	9.20	0.14	2.93	5.91
Argentina	17.44	0.95	1.10	1.88
Japan	0.37	–	1.18	1.34
Australia	19.70	6.24	1.55	2.78
USSR	4.44	0.62	2.78	1.18

Country	Edible offal protein (kg/cap.)	Egg protein (kg/cap.)	Net milk protein (kg/cap.)	Total animal protein (kg/cap.)
Canada	0.52	1.66	8.33	26.22
USA	1.09	2.22	6.80	28.29
Argentina	3.22	0.94	6.36	31.89
Japan	0.18	2.01	1.26	6.34
Australia	3.96	1.87	9.67	45.77
USSR	–	1.59	8.95	19.56

Table 38.

Country	Feed grain consumption (kg/capita)	Animal protein production (kg/capita)	Grain/protein ratio
Canada	737	26.22	28.1
USA	578	28.29	20.4
Argentina	388	31.89	12.2
Japan	183	6.34	28.9
Australia	434	45.77	9.5
USSR	536	19.56	27.4

hay from permanent pastures and grassland in rotation. This is shown by the high correlation between the feed grain-protein ration and the ratio of cereal area to equivalent agricultural area (Table 39, last column).

The equivalent agricultural area is the arable area plus one quarter of the permanent pasture area.* (The pasture area is divided by four as the mean grass yield is one quarter of the mean crop yield.)

Table 39.

Country	Cereal area (1974) (10^6 ha)	Arable area (1974) (10^6 ha)	Perm. pasture (1974) (10^6 ha)	Cereal area as percentage of equivalent agricultural area
Canada	17.87	43.1	25	36
USA	68.74	206.9	220	26
Argentina	11.43	34.4	144	16
Japan	2.94	5.6	0.24	52
Australia	11.60	45.0	454	7
USSR	121.41	232.7	375	37
Total	234	567.7	1218	27
World	725	1507	3044	32

Source: FAO Production Yearbook, 1975.

*This method of computing equivalent agricultural area is used by T. Kristensen in "Development in Rich and Poor Countries" (Praeger, New York, 1974).

III THE GLOBAL FOOD AND FEED BALANCE

The following is an estimate of world food and feed resources and their mode of utilization in 1974.

Resources

Table 40.

Land category	Type of product	Area (10^6 ha)	Yield (t grain-eqv./ha)	Production (10^6 t grain-eqv.)
Arable	Cereals and other field crops, plantation crops, horticultural products	1000	1.9	1900
Arable	Hay and other harvested roughage	300	1.0	300
Permanent pasture	Grass	3000	0.4	1200
Total		4300		3400

Utilization

The following Table shows global food availability in 1974.

Table 41.

Food	Energy content (kcal/capita/day)	Protein content (g/capita/day)
Vegetable products	2121	44.6
Animal products	447	24.4
of which: livestock products	~ 417	20.4
marine products	~ 30	4.0
Total food	2568	69.0

Source: FAO Production Yearbook, 1976.

World population in 1974 was 3.9 billions. It will be assumed that the overall conversion ratio of feedstuff energy to net available livestock product energy was 12 (the resultant of 35 for beef, mutton and goatmeat and 8 for milk, pigmeat, poultrymeat and eggs). It will also be assumed that seed requirements, industrial uses and storage losses accounted for 15% of the gross production of the arable area. One kg grain-equivalent is taken as 3320 kcal. The following estimates for utilization of agricultural products are based on these assumptions.

Table 42.

Utilization		Consumption (10^6 t grain-eqv.)
Food for humans	$2121 \times 365 \times 3.9 \times 10^9/3320 \times 10^3$ t	= 910
Feed for animals	$12 \times 417 \times 365 \times 3.9 \times 10^9/3320 \times 10^3$ t	= 2150
Seed, losses, etc.	$0.15 \times 2200 \times 10^6$ t	= 330
Total		= 3390

The food and feed balance in 2050

An optimistic assessment of food and feed resources in 2050 is given in the following Table.

Table 43.

Product	Area (10^6 ha)	Yield (t grain-eqv./ha)	Production (10^6 t grain-eqv.)
Field crops, etc.	1000	4.7	4700
Hay and other harvested roughage	400	2	800
Grass	3000	0.4	1200
Total	4400		6700

Assuming that:

(i) World population in 2050 will be 10 billions,

(ii) the seed requirement and storage losses in 2050 will be reduced to 10%,

(iii) the overall input-output ratio of feed to livestock products will be reduced to 9 (the current ratio in the EEC),

(iv) the global per capita intake of dietary energy remains constant at 2600 kcal,

(v) the energy content of livestock products is 18 kcal/g protein,

a global production of 6700 million tons grain-equivalent will provide a daily per capita intake of 21 g protein from livestock products.* If it is further assumed that the daily intake of protein of marine origin remains constant at 4 g, the total intake of animal protein will be 25 g, i.e. approximately the same as in 1974.

If the population in 2050 were 8 billions, the animal protein intake on assumptions (ii) to (v) would be 35 g. The population limit for an animal protein intake of at least 40 g is 7.2 billions, corresponding to 1000 per km^2 harvested cereal area, or 500 per km^2 arable area.

*If the daily intake of livestock product protein is x g, x is given by the relation

$$(2600 - 18x) + (9 \times 18x) = (0.9 \times 5500 + 1200)\, 10^9 \times 3320 \times (10^{10} \times 365)^{-1}$$

Appendix B

UTILIZATION OF THE ARABLE AREA

The global situation in 1976 as regards arable land and its utilization is summarized in Table 44. Column A gives the arable area, including permanent crops. Column B gives the total harvested area for the principal field crops (cereals, roots and tubers, soybeans, pulses, sugarbeet, sugar cane, groundnuts, sunflower seed, linseed, rapeseed and sesame seed).

Column C gives the harvested area for cereals.

Table 44.

	A Arable area (10^6 ha)	B Cropped area (10^6 ha)	C Grain area (10^6 ha)	B/A (%)	C/B (%)	C/A (%)
DC	695	378	310	54	82	45
LDC	811	620	449	76	72	55
World	1506	998	759	66	76	50
DC/World (%)	46	38	41			

Source: FAO Production Yearbook, 1976.

The distribution of the harvested areas for the major categories of crops in 1976 are given in Table 45.

Table 45.

Crop	DC area (10^3 ha)	LDC area (10^3 ha)	World area (10^3 ha)
Cereals	310408	449014	759422
Pulses	10120	60314	70434
Roots and tubers	14387	37166	51553
Soybeans	21344	23541	44885
Oilseeds*	11845	18699	30544
Sugar crops	9323	12785	22108
Groundnuts	896	18415	19311
Grapes	7742	2352	10094
Cocoa beans	–	4829	4829
Total	386065	627115	1013180

*Sunflower seed, sesame seed, linseed, rapeseed.

Source: FAO Production Yearbook, 1976.

The areas planted to non-food crops in 1976 are given in Table 46.

Table 46.

Crop	DC area (10^3 ha)	LDC area (10^3 ha)	World area (10^3 ha)
Seed cotton	7789	23574	31363
Coffee	1	7922	7923
Tobacco	1377	3100	4477
Jute and substitutes	16	2703	2719
Tea	137	1408	1545
Flax	1464	41	1505
Sisal	9	801	810
Hemp	229	200	429
Total	11022	39749	50771

Source: FAO Production Yearbook, 1976.

Table 47 gives the arable area, the harvested area for the principal field crops (the same categories as in Table 44) and the harvested cereal area for the eleven countries with the largest arable areas. The B/A ratios for China and Brazil are due to double cropping.

Table 47.

	A Arable area (10^3 ha)	B Cropped area (10^3ha)	C Grain area 10^3 ha)	B/A (%)	C/B (%)	C/A (%)
USSR	232207	145260	122554	63	84	53
USA	209236	96863	73306	46	76	35
India	167200	147803	104353	88	71	62
China	129000	173516	125152	134	72	97
Australia	45874	13692	12969	30	95	28
Canada	43767	21238	19674	49	93	45
Brazil	36600	37253	21536	102	58	59
Argentina	34550	16907	13523	49	80	39
Turkey	28286	14626	13056	52	89	46
Mexico	28000	12621	9541	45	76	34
Nigeria	23750	21855	12793	92	58	54

Source: FAO Production Yearbook, 1976.

Appendix C

EFFICIENCY OF PHOTOSYNTHESIS

I. QUANTUM EFFICIENCY

The theoretical maximum efficiency for the photosynthetic reduction of carbon dioxide to starch with white light can be calculated as follows:

Liberation of the carbon atom from a molecule of carbon dioxide requires eight photons of white light. The wavelength range for photosynthetically active radiation is 400 – 700 nm. The energy input in this range at the Earth's surface is equivalent to that of monochromatic light of wavelength 575 nm. (J. A. Bassham, "Increasing Crop Production Through More Controlled Photosynthesis," Science 197, 1977, p. 630-8).

8 photons of $\lambda = 575$ nm are equivalent to $8hc/\lambda$	$= 2.76 \times 10^{-11}$ erg
(h = Planck's constant, c = velocity of light)	
Mass of carbon atom	$= 1.99 \times 10^{-23}$ g
Energy required to liberate carbon	$= 1.39 \times 10^{12}$ erg/g
Energy content of starch (CH_2O)	= 114 kcal/mole
	= 3800 cal/g
	$= 1.59 \times 10^{11}$ erg/g
Carbon content of starch = C/CH_2O	= 0.40
Energy content of CH_2O per gramme C	$= 3.97 \times 10^{11}$ erg
Conversion efficiency = $3.97 \times 10^{11}/1.39 \times 10^{12}$	= 0.286

II LIMITING EFFICIENCY IN CEREAL PLANTS

The quantum efficiency of photosynthesis cannot be attained in crop plants because of respiration losses (return of CO_2 to the atmosphere), reflection of light from the leaf surfaces, absorption of energy by inactive tissues and light saturation in the upper leaves of the plant.

The magnitudes of these losses during the grain-filling period for rice have been given by Y. Murata and S. Matsushima, and for cereals in general by L. T. Evans. The values are tabulated below. Evans distinguishes between C_3 plants (wheat, rice) and C_4 plants (maize, sorghum); the starch molecules of the former contain three carbon atoms, of the latter four.

The light saturation loss given by Murata and Matsushima is based on a total incident radiation of 400 cal/cm^2/day; Evans' figures refer to an incident radiation of 500 cal/cm^2/day.

Both Murata and Matsushima and Evans agree that the daily respiration loss is 25% of gross photosynthesis plus 1.5% of the accumulated dry weight of plant material at the start of the grain-filling period. For the following calculation the latter will be taken as 1500 g/m^2, as assumed by Murata and Matsushima.

The maximum overall efficiency of utilization of photosynthetically active radiation (PAR) in cereal plants is calculated below on the basis

Table 48.

Loss	Murata and Matsushima	Evans	
	(Rice)	(C_3 plants)	(C_4 plants)
Reflection (%)	5.5	16	16
Absorption (%)	10	10	10
Light saturation (%)	17	36	12

Sources: Y. Murata and S. Matsushima, "Rice", a contribution to the Symposium "Crop Physiology. Some Case Histories", (ed. L. T. Evans), Cambridge University Press, 1975. L. T. Evans, "The effect of light on plant growth, development and yield", a contribution to the Symposium "Plant Response to Climatic Factors", UNESCO, 1973.

of the losses given above. It is assumed that 45% of the total incident radiation is photosynthetically active, that the quantum efficiency of photosynthesis is 0.286, and that the energy content of dry plant material is 3800 cal/g.

Murata and Matsushima: Rice

$$PAR = 400 \times 10^4 \times 0.45 = 1.8 \times 10^6 \text{ cal/m}^2\text{/day}$$

Maximum rate of PAR conversion

$$= (1.8 \times 10^6 \times 0.286 \times 0.945 \times 0.9 \times 0.83 \times 0.75 - 1500 \times 0.015 \times 3800)$$

$$= 1.88 \times 10^5 \text{ cal/m}^2\text{/day}$$

Maximum photosynthetic efficiency $=1.88 \times 10^5/1.8 \times 10^6 = 0.10$.

Evans: Cereals

$$PAR = 500 \times 10^4 \times 0.45 = 2.25 \times 10^6 \text{ cal/m}^2\text{/day.}$$

Maximum rate of PAR conversion in C_3 plants

$$= (2.25 \times 10^6 \times 0.286 \times 0.84 \times 0.9 \times 0.64 \times 0.75 - 1500 \times 0.015 \times 3800)$$

$$= 1.48 \times 10^5 \text{ cal/m}^2\text{/day.}$$

Maximum photosynthetic efficiency $= 1.48 \times 10^5/2.25 \times 10^6 = 0.07$

Maximum rate of PAR conversion in C_4 plants

$$= (2.25 \times 10^6 \times 0.286 \times 0.84 \times 0.9 \times 0.88 \times 0.75 - 1500 \times 0.015 \times 3800)$$

$$= 2.35 \times 10^5 \text{ cal/m}^2\text{/day.}$$

Maximum photosynthetic efficiency $= 2.35 \times 10^5/2.25 \times 10^6 = 0.10$.

Appendix D

GRAIN YIELD AND CHEMICAL NITROGEN FERTILIZER

The relation between grain yield and intensity of application of chemical nitrogen fertilizer can be determined on a world basis as follows. The calculation is based on data from FAO Production Yearbooks 1973, 1974 and 1975.

The mean intensity of nitrogen application is found by dividing total nitrogen consumption by the sum of the harvested areas for cereals, roots and tubers, soybeans, pulses, sugarbeet, sugarcane, oilseeds and groundnuts.

X_1 = mean intensity of nitrogen application 1961-5	= 16.3 kg/ha
X_2 = mean intensity of nitrogen application 1972-4	= 38.1 kg/ha
$\Delta X = X_2 - X_1$	= 21.8 kg/ha
Y_1 =mean world grain yield 1961-5	= 1460 kg/ha
Y_2 = mean world grain yield 1971-5	= 1848 kg/ha
$\Delta Y = Y_2 - Y_1$	= 388 kg/ha

The incremental yield/nitrogen is thus

$\Delta Y/\Delta X$	=17.8

Y_0, the yield corresponding to $X = 0$, is

$$Y_2 - [(\Delta Y/\Delta X) \times X_2)] = 1171 \text{ kg/ha}$$

The relation between X and Y is then given by

$$Y = 1171 + 17.8X.$$

Values of $\Delta Y/\Delta X$ and Y_0 for seven major regions, and the data on which they are based, are given in Table 49.

Table 49.

Region	Y_1 (kg/ha)	Y_2 (kg/ha)	X_1 (kg/ha)	X_2 (kg/ha)	$\Delta Y/\Delta X$	Y_0 (kg/ha)
Developed market economies	2342	3028	49.7	87.3	18.3	1432
USSR and East Europe	1173	1724	14.3	53.2	14.2	971
Latin America	1331	1561	9.4	22.9	17.0	1169
Africa	742	794	1.5	4.8	15.7	719
Near East	1128	1293	9.3	26.6	9.5	1041
Far East	1145	1388	4.6	15.6	22.0	1044
Asian centrally planned economies	1555	1871	7.0	22.7	20.1	1415

Appendix E

CARBON DIOXIDE AND THE ATMOSPHERE

The carbon content of all fossil fuels burned in 1975 is calculated in the Table below.

Table 50.

Type of fuel	Consumption (10^6 t coal-eqv.)	Carbon/energy ratio (t/t coal-eqv.)	Carbon content (10^6 t)
Solid fuels	2623	0.85	2230
Liquid fuels	3526	0.65	2290
Natural gas	1633	0.45	735
Total	7782	0.68	5255

Molecular weight ratio of CO_2 to C = 44/12 = 3.67

Global CO_2 emission in 1975 = $5.255 \times 10^9 \times 3.67 = 1.93 \times 10^{10}$ t

Mass of the atmosphere = $(4\pi R^2 h) \times 10^9 = 5.26 \times 10^{15}$ t

(where R = mean radius of the Earth = 6370 km and h = hydrostatic equivalent atmospheric pressure = 0.0103 km).

Mass ratio of CO_2 emission to the atmosphere = 3.66 ppm.

Molecular weight ratio of air to CO_2 =28.8/44 = 0.65

Volumetric ratio of emission to atmosphere = 3.66 × 0.65 = 2.4 ppm

Atmospheric CO_2 concentration in 1975 = 327 ppm

Observed annual increase of CO_2 concentration 1970-5 = 1.2 ppm

Proportion of global emission remaining in the atmosphere=50%

Assuming that the proportion of the global emission remaining in the atmosphere does not change, that fossil fuel consumption increases by 4% annually, and that the mean carbon-energy ratio of all fossil fuels remains at the 1975 figure, the atmospheric CO_2 concentration in the year 2000 will be

$$327 + 1.2\ \frac{1 - 1.04^{25}}{1 - 1.04} = 379\ \text{ppm}.$$

Recent Papers on atmospheric carbon dioxide

C. F. Baes *et al.*, "Carbon Dioxide and the Climate: The Uncontrolled Experiment", *American Scientist*, May-June 1977, 310-320.

W. W. Kellogg, "Effects of Human Activities on Global Climate", *Bulletin of World Meteorological Organization*, Oct. 1977, pp. 229-240; Jan. 1978, 3-10.

M. Stuiver, "Atmospheric Carbon Dioxide and Carbon Reservoir Changes", *Science* **199** (1978), 253-258.

U. Siegenthaler and H. Oeschger, "Predicting Future Atmospheric Carbon Dioxide Levels", *Science*, **199** (1978), 388-395.

C. F. Cooper, "What Might Man-Induced Climate Change Mean?", *Foreign Affairs*, Apr. 1978, 500-520.

Appendix F

THE FUSION ENERGY POTENTIAL OF THE HYDROSPHERE

The four fusion reactions are:

$$_1D^2 + {}_1D^2 \rightarrow {}_2He^3 + {}_0n^1 + 3.27 \text{ MeV}$$

$$_1D^2 + {}_1D^2 \rightarrow {}_1T^3 + {}_1H^1 + 4.03 \text{ MeV}$$

$$_1D^2 + {}_1T^3 \rightarrow {}_2He^4 + {}_0n^1 + 17.6 \text{ MeV}$$

$$_1D^2 + {}_2He^3 \rightarrow {}_2He^4 + {}_1H^1 + 18.3 \text{ MeV}$$

Assuming that the technical problems involved in maintaining the four reactions simultaneously will ultimately be solved, the net reaction will be:

$$6\,{}_1D^2 \rightarrow 2\,{}_2He^4 + 2\,{}_1H^1 + 2\,{}_0n^1 + 43.2 \text{ MeV}$$

i.e. six atoms of deuterium will yield 43.2 MeV of energy.

The atomic weights of the particles involved are:

Deuterium	$_1D^2$	2.01410
Helium	$_2He^4$	4.00260
Hydrogen	$_1H^1$	1.00782
Neutron	$_0n^1$	1.00867

$$6\,{}_1D^2 = 12.0846 \text{ atomic mass units}$$

$$2\,{}_2He^4 + 2\,{}_1H^1 + 2\,{}_0n^1 = 12.0382 \text{ atomic mass units}$$

It follows that

$$(12.0846 - 12.0382)/12.0846 = 0.00384$$

of the deuterium is converted to energy.

Seawater contains 10.67% by weight of hydrogen. Hydrogen contains 0.0298% by weight of deuterium. One ton of seawater therefore contains 31.8 g deuterium, of which

$$0.00384 \times 31.8 = 0.122 \text{ g}$$

will be converted to energy. As energy is the product of mass and the square of the velocity of light ($E = mc^2$) it follows that one ton of seawater will yield

$$0.122\,(3 \times 10^{10})^2 = 1.1 \times 10^{20} \text{ ergs}$$
$$= 1.1 \times 10^{13} \text{ joules}$$
$$= 380 \text{ t coal-equivalent.}$$

The volume of the hydrosphere = 1.3×10^{18} m^3.

The fusion energy potential of the deuterium of the hydrosphere is therefore

$$1.3 \times 10^{18} \times 380 = 4.9 \times 10^{20} \text{ t coal -equivalent}$$

Note: The energy required to separate deuterium from water by distillation is approx. 2000 kWh/kg D_2O, corresponding to 10 000 kWh/kg D_2. ("Encyclopedia of Electrochemistry", Reinhold, New York, 1964).

The energy needed to separate the deuterium from one tonne of seawater is therefore approximately

$$10^4 \times 0.0318 = 318 \text{ kWh} \approx 0.04 \text{ t coal equivalent.}$$

As the theoretical energy output of the fusion reactions is 380 t coal equivalent/m^3 seawater, the output/input ratio is

$$380/0.04 \approx 10\,000.$$

The present state of fusion research and the outlook for utilization of fusion energy is discussed by J. P. Holdren in "Fusion Energy in Context: Its Fitness for the Long Term" (*Science*, **200**, 168-180, 14 April, 1978). He concludes that fusion energy will probably make no substantial contribution to the world energy supply until at least two decades into the next century – and it is likely to be expensive.

Appendix G

FOOD AVAILABILITY – FAO AND OECD ESTIMATES

The tabulated data are for 1973 except where otherwise stated.

Table 51.

Country	FAO estimate			OECD estimate		
	Energy (kcal)	Total protein (g)	Animal protein (g)	Energy (kcal)	Total protein (g)	Animal protein (g)
Australia*	3383	101.0	67.3	3048	100.8	69.3
Austria	3516	87.6	51.7	3332	87.8	53.9
Belgium-Luxemb.	3699	99.0	58.9	3441	96.3	58.6
Canada	3410	98.6	64.4	3158	95.9	65.0
Denmark	3452	89.2	56.9	3229	92.2	65.5
Finland*	3222	91.8	59.1	3126	96.4	66.5
France	3425	98.0	61.3	3219	102.7	67.0
German Fed. Rep.	3456	86.8	54.6	3238	88.0	57.5
Ireland	3544	102.6	62.0	3379	100.5	62.3
Italy	3568	100.1	43.4	3343	105.3	48.9
Japan	2850	87.4	42.2	2571	81.2	32.2
Netherlands	3304	86.0	53.2	3176	88.8	57.9
New Zealand*	3492	106.7	72.8	3232	106.7	73.6
Norway	3203	90.1	58.1	3011	85.0	54.6
Portugal	3426	92.2	38.2	3297	93.2	39.0
Spain	3228	90.6	44.6	2829	90.2	45.9
Sweden	3022	85.1	55.6	2760	83.9	57.6
Switzerland*	3543	90.4	56.4	3296	92.5	57.5
UK	3351	92.0	55.8	3130	87.9	54.6
USA	3553	103.7	70.6	3316	103.6	72.6
Yugoslavia	3349	93.2	28.5	3164	92.9	28.6

*1972.
Sources: FAO Production Yearbook, 1976. OECD Food Consumption Statistics 1955–1973.

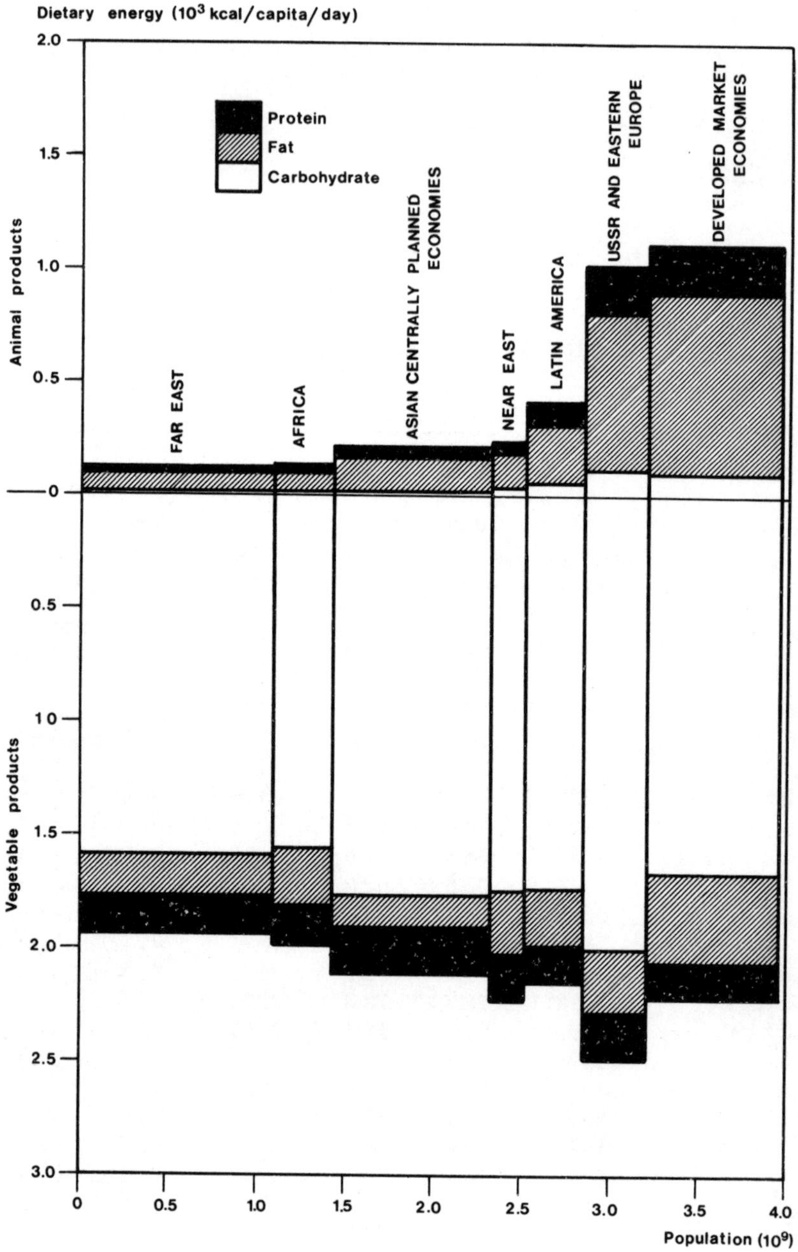

Fig. 20. Food availability by regions, 1974. (Source: FAO Production Yearbook, 1976.);

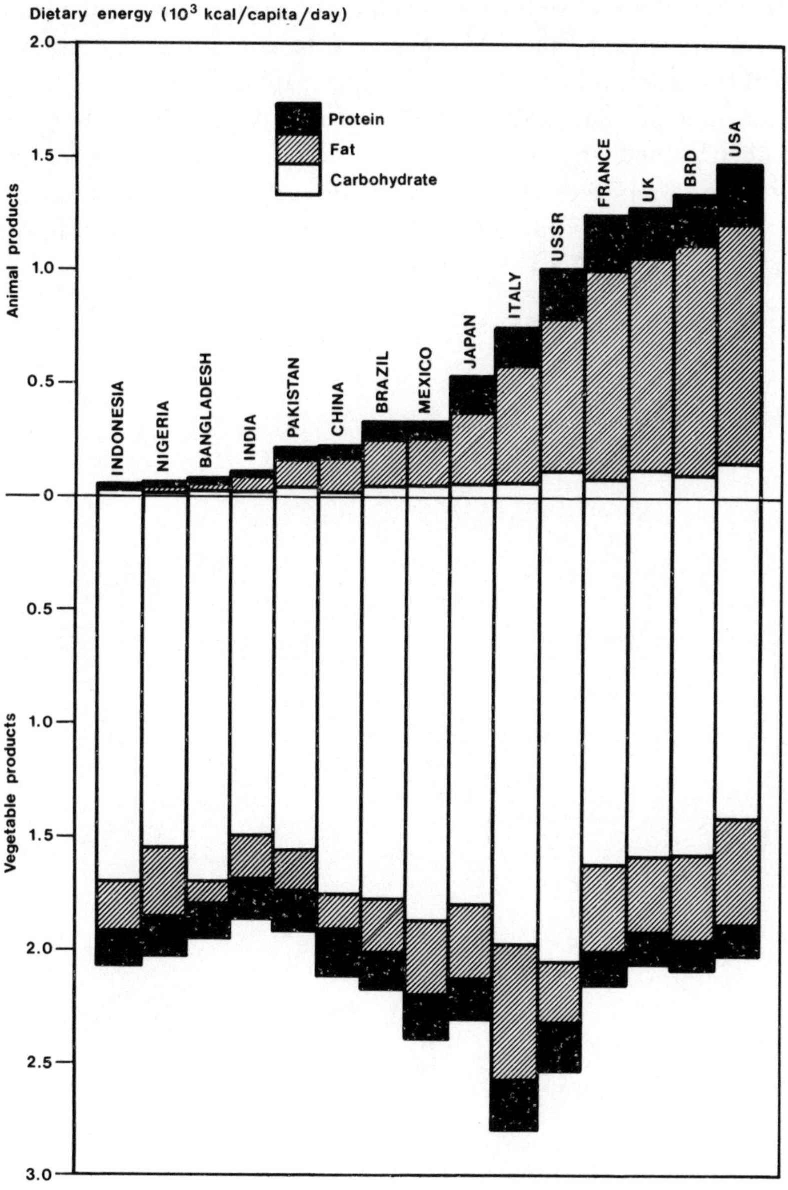

Fig. 21. Food availability by countries, 1974. (Source FAO Production Yearbook, 1976).

The only estimates of food availability for countries other than OECD members are those of the FAO. The most recent estimates are shown in Figs. 20 and 21. The former shows per capita availability of animal protein, animal fat, animal carbohydrate, vegetable carbohydrate, vegetable fat and vegetable protein in each of seven major regions in 1974. The latter gives the same data for each of the fifteen countries with populations exceeding 50 millions. The energy content of protein and carbohydrate is 3.8 kcal/g, of fat 8.8 kcal/g.

Appendix H

POPULATION PROJECTIONS

Table 52 shows the projected populations for the year 2000 for each of the eight major areas, as computed by the UN demographers. All figures are in millions.

Table 52.

Area	Low variant	Medium variant	High variant
N. America	279	296	334
Europe	519	540	561
USSR	305	315	324
Oceania	30	33	35
Latin America	569	620	671
Africa	747	814	856
South Asia	2103	2267	2421
East Asia	1287	1370	1436
World	5840	6254	6638

Source: "World Population Prospects as Assessed in 1973", UN Population Study No. 60, New York, 1977.

The UN projections can be compared with those computed for the same year by Dr. T. Frejka, given in Table 53.

The annual UN estimates of world population are of limited value for assessing the probabilities of the three UN projections on a global basis, as in these estimates it is *assumed* that the populations of several major LDCs (China, Indonesia, Nigeria) are following the medium variant projection. An indication of the trend of demographic development can, however, be obtained by comparing the projected annual rates of population growth for the period 1975-80 with the estimated growth in the period mid-1976 to mid-1977. These statistics are shown for selected countries in Fig.22.

Table 53.

Area	Proj. 1	Proj. 2	Proj. 3	Proj. 4	Proj. 5
N. America	277	286	302	310	315
Europe	527	540	561	572	578
USSR	302	307	314	318	320
Oceania	23	25	27	28	29
Latin America	405	447	540	598	626
Africa	464	510	620	693	729
South Asia	1522	1687	2079	2339	2469
East Asia	1229	1314	1485	1592	1646
World	4749	5116	5928	6450	6712

Source: T. Frejka, "The Future of Population Growth", New York, 1973.

NB! The annual population growth in Mexico for the period 1975-80 on the UN low variant projection is 3.26%, and not 2.36% as shown in Fig. 22. The incorrect figure is tabulated in UN Population Study No. 60.

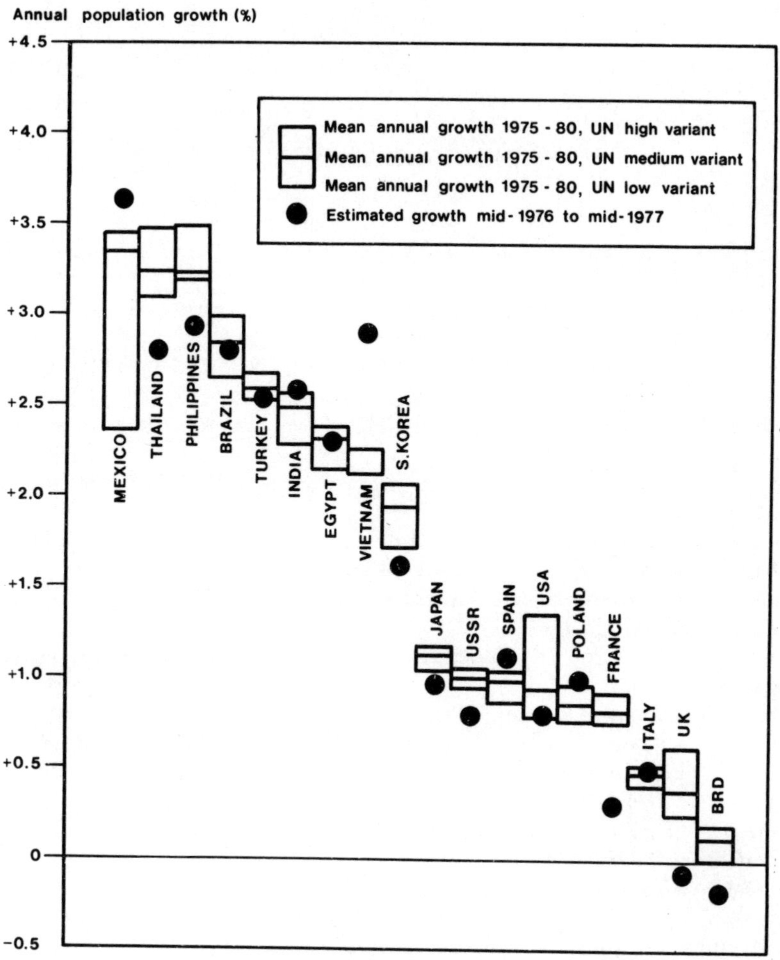

Fig. 22. Current population growth. (Sources: "World Population Prospects as Assessed in 1973," UN, New York, 1977. UN Monthly Bulletin of Statistics, May 1978.)

BIBLIOGRAPHY

Borlaug, N. E. "The Destiny of Man and World Civilization." University of Arkansas, Little Rock, 1974.

Brown, H., Bonner, J. and Weir, J. "The Next Hundred Years". Viking, New York, 1957.

Brown, L. R. "In the Human Interest: A Strategy to Stabilize World Population". Norton, New York, 1974.

Darwin, C. G. "The Next Million Years". Hart-Davis, London, 1952. Repr. Greenwood Press, Westport, Connecticut, 1973.

Davis, K. "Population Policy: Will Current Programs Succeed?" *Science* **158**, 10 Nov. 1967.

Echols, J. R. "Population vs. the Environment: A Crisis of Too Many People." *American Scientist* **64**, Mar. 1976, 165-173.

Eckholm, E. P. "Losing Ground: Environmental Stress and World Food Prospects". Norton, New York, 1976.

Evans, L. T. (ed.) "Crop Physiology: Some Case Histories". Cambridge University Press, 1975.

Foley, G. "The Energy Question". Penguin, London, 1976.

Fraser, D. "The People Problem". Indiana University Press, Bloomington, 1971.

Frejka, T. "The Future of Population Growth: Alternative Paths to Equilibrium". Wiley, New York, 1973.

Fukuda, H. "Irrigation in the World: Comparative Developments." University of Tokyo Press, 1976.

Grist, D. H. "Rice". Longman, London, 5th ed. 1975.

Hutchinson, J. (ed.) "Population and Food Supply: Essays on Human Needs and Agricultural Prospects". Cambridge University Press, 1969.

Janzen, D. H. "Additional Land at What Price? – Responsible Use of the Tropics in a Food-Population Confrontation." *Proc. Amer. Phytopathol. Soc.* **3**, 1976, 35-39.

Jensen, N. F. "Limits to Growth in World Food Production: Ceilings for wheat yields are coming in developed countries." *Science* **201**, 28 July 1978, 317-320.

Kamarck, A. M. "The Tropics and Economic Development: A Provocative Inquiry into the Poverty of Nations." Johns Hopkins University Press, Baltimore, 1976.

Kenward, M. "Potential Energy: An Analysis of World Energy Technology." Cambridge University Press, 1976.

Kristensen, T. "Development in Rich and Poor Countries". Praeger, New York, 1974.

Kuznets, S. "Population, Capital and Growth." Norton, New York, 1973.

Laqueur, W. "Third World Fantasies." *Commentary* **63**, No. 2, Feb. 1977.

Lehner, G. "Possibilities of utilisation of new non-nuclear and non-fossil energy sources." *Kerntechnik* **20**, 1978, 157-168.

Loomis, R. S., Williams, W. A., and Hall, A. E. "Agricultural productivity." *Annual Rev. Plant Physiol.* **22**, Palo Alto, 1971, 431-468.

McCarrison, R., Sinclair, H. M. "Nutrition and Health". Faber, London, 3rd ed. 1961.

McMullan, J. T., Morgan, R. and Murray, R. B. "Energy Resources and Supply". Wiley, London, 1976.

McNamara, R. S. "Address to the Massachusetts Institute of Technology on the Population Problem." International Bank for Reconstruction and Development, Washington DC, 1977.

National Research Council. "Agricultural Production Efficiency". Nat. Acad. Sci., Washington DC, 1975.

National Research Council. "World Food and Nutrition Study: The Potential Contributions of Research." Nat. Acad. Sci., Washington DC, 1977.

Paddock, W. C. "A Humanitarian Response by Scientists to the Food/Population Equation: A Moratorium on Agricultural Research and a 'Watchdog Committee' to Monitor Claims of Agricultural Breakthroughs." *Proc. Amer. Phytopathol. Soc.* **3**, 1976, 40-46.

Parsons, J. "Population Fallacies". Elek-Pemberton, London, 1977.

Pimentel, D., *et al.* "Land Degradation: Effects on Food and Energy Resources." *Science* **194**, 8 Oct. 1976.

President's Science Advisory Committee. "The World Food Problem". White House, Washington DC, 1967.

Scrimshaw, N. S., Young, V. R. "The Requirements of Human Nutrition." *Scientific American* **235**, Sept. 1976.

Smil, V. "Food in China." *Current History* **75**, Sept. 1978, 69-72, 82-84.

Teitelbaum, M. S. "Relevance of Demographic Transition Theory for Developing Countries." *Science* **188**, 2 May 1975.

Thompson, J. D. "Ocean Deserts and Ocean Oases." *Climatic Change* **1**, 1978, 205-230.

Tont, S. A., Delistraty, D. A. "Food resources of the Oceans: An Outline of Status and Potentials." *Environmental Conservation* **4**, 1977, 243-252.

United Nations Department Of Economic And Social Affairs. "Concise Report on the World Population Situation in 1970-75 and its Long-Range Implications". Population Study No. 56, New York, 1974.

United Nations Department Of Economic And Social Affairs. "World Population Prospects as Assessed in 1973". Population Study No. 60. New York, 1977.

Vogt, W. "People! Challenge to Survival". Gollancz, London, 1961.

Wareing, P. F., Cooper, J. P. (eds.) "Potential Crop Production: A Case Study". Heinemann, London, 1971.

Weinberg, A. M. "Reflections on the Energy Wars." *American Scientist* **66**, Mar. 1978, 153-158.

World Energy Conference. "Survey of Energy Resources". U. S. National Committee of the W. E. C., New York, 1974.

INDEX

F

G

H

I

K

L

M

N

O

P